Thomas H Walton

Coal Mining Described and Illustrated

Thomas H Walton

Coal Mining Described and Illustrated

ISBN/EAN: 9783743320048

Manufactured in Europe, USA, Canada, Australia, Japa

Cover: Foto ©ninafisch / pixelio.de

Manufactured and distributed by brebook publishing software (www.brebook.com)

Thomas H Walton

Coal Mining Described and Illustrated

PREFACE.

The Almighty in His great bounty has provided for all our wants in endless abundance and variety. In some parts of the Tropics, when the rainy seasons come to moisten the ground which has been baked by the heat of the sun, you can see the vegetation thrust out of it by a powerful influence. The transformation which takes place is truly magical. Within a few weeks, the ground that was bare supports masses of vegetable matter so high and so rank as to forbid one's passage through them. Year succeeds year and each contributes its layer of vegetation, which is generously supplied with carbon, oxygen, hydrogen, nitrogen, and traces of other important ingredients which combine to form those wonderful products of nature.

But, alas! after the rains cease to moisten the earth, the sun shines hotly through the thirsty atmosphere and abstracts the moisture from the vegetation which has been so generously nourished by it. In a short time only a portion of the whole remains; the bulk of the ingredients passes back into the winds again, and is carried by them to other parts of the globe perhaps to make component parts in the construction of other vegetable or animal organisms.

But all the species of the vegetable kingdom are not so frail in nature. Where the ground is liberally supplied with moisture from subterranean sources, the huge trunks of trees spread their branches into the air, and those branches carry their foliage so compactly as to defy the rays of the sun, and forbid its stealing the moisture from the ground which is nourishing the roots. In this case the vegetable accumulations augment, and continue to do so year after year as long as they live. A step farther on—a geological step, which is in all cases a long stride—we find the vegetable masses dead and buried deeply in the earth, imbedded in their hard rocky graves! Such in fact are our coal seams.

These vegetable preservations of the past are handed down to us in the shape of fuel; and after we use them as such, the bulk of the ingredients which have been held prisoners for ages, is again set at liberty and takes its place in the air ready for a new commencement. After so much pains have been taken in the formation of these coal seams, it is wantonness to waste their contents by bad mining operations.

The object of this work is to point out the present modes of taking out the coal contained in coal seams, as practised in England and the United States.

viii   PREFACE.

To extract coal from seams lying on low inclinations has been very successfully practised in the older mining countries. In some of the provinces of France, where the coal seams incline upwards of 70°, and which are comparatively thick, much difficulty was formerly experienced in attempts made to mine out the coal entirely. Galleries driven into the coal seams horizontally, with pillars of certain widths left between them, resulted in much of the coal being lost, after it was detached from the pillars. It was then often buried by the rocks which caved in near the ends of these pillars. This occasioned a change to the present system of working by *remblais*, which seems to be a plan well adapted to the taking out of the coal of highly inclined coal seams.

Ventilation of coal mines is treated in the following pages only in the most practical manner. To comprehend the fine points connected with mine ventilation on an extensive scale, an accurate knowledge of pneumatics with much practical experience is absolutely necessary.

Absurdities connected with this important subject have floated into the works of very eminent writers, and it will not be an easy task to eradicate them.

The chief difficulties in the way of practical ventilation of mines seem to consist in the maintenance of the air-passages of sufficient size free from obstructions, and in the resistance met with by air-currents moving with high velocities.

Trusting that the succeeding pages will be found of use to the student of coal mining and to the operators of coal mines and the owners of coal lands, and that they may be of interest to the general reader as well, the author passes them over to the public with a consciousness of having thus done something towards performing his duty towards his fellow man.

PHILADELPHIA, December 15, 1884.

CONTENTS.

SECTION I.

GENERAL INFORMATION CONCERNING COAL MINES AND COAL MINERS, WITH DESCRIPTION OF LONG WALL AS WORKED IN HORIZONTAL SEAMS OF COAL IN ENGLAND AND IN FRANCE.

CHAPTER I.

EXPLOSION AT SPRINGWELL COLLIERY. MEANS TO BE USED AS SAFEGUARDS AGAINST ACCIDENTS.

	PAGE
Personal Recollections of an Explosion at Springwell Colliery	9
Improvements introduced at Coal Mines as a Safeguard against Accidents; Practical details .	12

CHAPTER II.

COAL-DUST AND COAL-GAS IN THE AIR FORMING A DANGEROUS MIXTURE.

Large quantities of gas developed in Coal Mining; What an explosive atmosphere may be composed of; Coal-dust finely pulverized disseminated through the air, a dangerous constituent; Mode of working thick beds of Anthracite which lie on a high inclination . . . 14

CHAPTER III.

MARSH GAS, AN EXPERIMENT. THE EFFECT OF A BRATTICE IN AN OPENING SCHUTE.

Marsh Gas identical with the Carburetted Hydrogen of a Coal Mine; Incidents connected with an experiment with this Gas	17
How Marsh Gas may be obtained for experimental purposes; An incident illustrating **prejudices** among Miners; A Colliery near Ashland, Pa.	19
Fan Boys and their Work; Dispensing with Fan Boys in getting rid of Gas . . .	20
The use of a piece of Canvas and a Brattice in Ventilation	21

CHAPTER IV.

IMPROVEMENTS OF MINING IN THE SOUTHERN AND WELSH MINES. **COAL MINING ENGINEERS.** HETTON COLLIERY.

Mining in the Lancashire District, England; Improvements introduced from the North . .	24
System of Organization of Labor at the Hetton Collieries in the North of England . .	25
Effects of the introduction of this System into Lancashire and Wales . . .	26
Details of this System as adopted in Lancashire	27
Principles which should govern a System of Ventilation; Necessity of recognizing a relation between the amount of Coal Mined and the amount of Air required in Ventilation . .	28

(ix)

CHAPTER V.

THICK COAL SEAMS—WORKING OUT COAL IN FRANCE BY REMBLAIS—IN ENGLAND BY LONG WALL.

SECTION II.

AN EXAMPLE OF MINING OUT COAL BY WHAT IS TERMED THE BOARD AND PILLAR SYSTEM.

CHAPTER VI.

SHAFT THROUGH ENGLISH COAL MEASURES—WINDING IN SHAFT—ENGINE PLANE—HORSE ROADS.

	PAGE
Board and Pillar System as practised in the Hutton Seam in the New Castle, England, Coal Field ; Description of the Vein and the Works	33

CHAPTER VII.

DISTRICT AND PANEL WORKINGS—BOARDS AND ENDS OF COAL—WORKING LEVELS—DISTRICT DETAILS.

Description of a Plate illustrating an Overman's Tracing ; Description of a Plate illustrating Plan of Board and Pillar Whole Coal Workings	36
The Board and Pillar or Board and Wall System, one of the most popular methods of Mining ; Origin of these Terms and their relation to the Cleavage of the coal ; Descriptive details of this System	37
The Aircrossing and the General System of Ventilation of Coal Mines by Splitting the main current of Air, as Illustrated by a Drawing	39
The Regulator in Ventilation, with a Description of its Mode of Operation	40
On ascertaining the quantity of Gas in the Air and the Tests used for that purpose ; Names of the places in the District as illustrated in the Plan of Board and Pillar Whole Coal Workings	41

CHAPTER VIII.

A DEPUTY'S EXPERIENCE—DETAILS IN WORKING AND VENTILATING A DISTRICT.

The Deputy ; An interview with him ; His views on the subject of Ventilation	42
Anemometer ; Practical methods of estimating the condition of the Air-current in a Mine	43
The use of the Water Gauge	45

CHAPTER IX.

THE PUTTER.

The Putter and his Duties	47
The Blackboard, showing the Miners' places, the number of Wagons of Coal to be cut in each place, and the Work of the Putters	48
Accidents to the Putters	49
Drawing the Cavils in order to determine the Stations of the Putters	50

CHAPTER X.

THE HEWER AND HIS WORK.

The Coal Miner Nicking his Jud	54
The Headways	55
The requirements demanded of the Miner in the New Castle Coal Field; The work of the Deputy in that Coal Field	56

CHAPTER XI.

THE OVERMAN—SELF-ACTING INCLINE PLANE.

The Overman, the responsibility of his Position and his Duties; The Master-waste Man and the Master Shifter and their Duties; An interview with the Overman	58
The Incline Bank	61

CHAPTER XII.

"BROKEN COAL" WORKED.

The various methods of taking out Pillars of Coal; The Miner's first step; The Trapper and his Duties; Air-doors; The Regulator and Air-crossing as a Substitute for Air-doors . .	63
Working the "Whole Coal"; The Tramway used in Durham and Northumberland . .	64
Blackboard showing the number of Tubs to each Jud and the Sheaths or Ranks of Putters .	65
Explanations by the Overman concerning the working off of the Juds, including the dimensions of the Pillars of a District, the time required to work off a Jud, the yield of Coal from a District working off the Broken Coal and the Crush of the Roof	66
The Drawing out of the Props	67
Fall of the Roof; The Deputies employed in a District	68
The Wagon-way man; Skelping the Coal; The use of the Safety Lamp; The yield of Gas; Keeping the Air safe from admixtures of Dust; Effect of Gas from an upper or lower Coal Seam forcing its way into the Goaf of a working Mine	69

CHAPTER XIII.

DETAILS OF BROKEN WORKINGS.

Method of working out the Pillars, when the Dip is great; Dangers of Explosions in following up the workings in the Whole Coal by workings in the "Broken" . .	71
Effect of a Goaf behind us, advancing and extending itself; The Long Wall System of Mining Coal	72

CHAPTER XIV.

REMARKS AND COMPARISONS.

Chief merit of the Plan of working shown in which the "Broken" District is worked simultaneously with a whole District which it follows at what is judged a safe distance; Difficulty of Ventilating a Goaf; Creeping down of the Roof	73

CONTENTS.

	PAGE
Respective advantages, under different circumstances, of the Board and Pillar and the Long Wall Systems of Mining	74
Mining in America; Advantages of taking out Pillars by working back in the direction of the Levels beginning at the Dip, or with these Levels driven on the lowest part of the boundary	75
Splitting the Pillars; Condition in which the Pillars and Excavations surrounding them are generally found; Metal Ridge or Rig; Tunnelling through the Metal Rigs; Source of trouble and expense of these Rigs, and manner of working through them	76
Closing up of the Board-rooms in working forward toward the boundary of the Mines	77

CHAPTER XV.

RE-WORKING OF OLD MINES, "METAL RIGS," AND OLD COAL PILLARS.

The mode of opening work through old Board-rooms closed up by Ridges, lifted from the Bottom Slate; Practical details	78
Driving Headways in the Whole Coal	79
Modification of Board and Pillar System still in general use in Lancashire, England	80

CHAPTER XVI.

GENERAL REMARKS.

The Overman's Cabin; Familiar chat with the Overman concerning the Men and the Mines of the Newcastle Coal Fields; Foundation of Railways here laid; Some of the older Mining appliances	81
Misfortunes of the past from a want of knowledge of Gases in the Mines	82
Terrible Explosions in the past history of Mining; Little knowledge of the means of Ventilation; Improvements of Spedding and Buddle in Ventilation	83
Advances in Mining at the present time; The Putter and his work	84
The "Calling Course"	85
The duties of the Caller	86
The Spare Hoisting Shaft	87
The interior of a Board-room, manner of Branching of the Road into it, and Mode of Propping up a Slate or Shaly Roof	88

SECTION III.

HOW COAL IS TAKEN FROM THE HIGHLY INCLINING COAL VEINS OF THE UNITED STATES.

CHAPTER XVII.

TOPOGRAPHICAL FEATURES—CHARACTERISTICS OF THE MINES.

Description of an extensive Coal Mine and its surroundings, in Schuylkill County, Pa.	89
View in the Gap or Ravine Cutting through a mountain in this localay; Interviews with Mine Bosses in an attempt to find employment	90
Gangway Timbers illustrated	91

CONTENTS. xiii

CHAPTER XVIII.
COAL FORMATIONS—DEPOSITS—UPHEAVALS.

	PAGE
The nature of Coal Strata examined	98
Characteristics of Coal Veins in Schuylkill County, Pennsylvania	99

CHAPTER XIX.
MINING OF COAL—MINERS' TOOLS—STARTING THE SCHUTES—DRILLING AND BLASTING.

The Miner's Outfit; The Mouth of the Drift; The Water Course in a Mine; Lagging and Loose Packing; A Timbered and Lagged Gangway 102
A Plan of the Coal Mines as they are worked in the Thick Coal Seams of the United States; Mines in the United States leased by the Ton and not by the Acre, as in other countries; Disadvantages of this System; Immense destruction and waste of Coal 103
A Schute and the manner of Driving it; The manner of connecting together the main Gangway and the Breast-way, and other details of Mining 104
The "Starter" at work; The Battery and the Battery Collar 105
A Schute shown "blocked up" at the Battery; The Travelling Road in a Mine; "Starting" a Battery; Description of a set of tools used to Mine Coal in the Breasts of the highly inclined Coal Beds of Pennsylvania; The Drill; The Needle 106
The Scraper; The mode of inserting the charge of powder in Blasting Coal . . . 107
The Squib 108
The style of Pick used in the Anthracite Coal region; The steel Sledge and the steel Wedge; Drilling 109
The Loader and the Driver and the Boss Loader 110

CHAPTER XX.
DRIVING A BREAST—COST OF COAL—MANWAY AND HEADING—BLOWING DOWN TOP COAL.

An Air-door; Mode of Timbering the Manways and Breast-rooms, and of connecting them together, by means of Headways driven through the Pillars, with other details . . 113
Breast in the Mammoth Vein illustrated 115

CHAPTER XXI.
VENTILATION—DRAINAGE OF WATER-LEVEL—GANGWAY IN BOTTOM ROCK.

Ventilation details; The pressure of the Atmosphere 120
Practical discussions 121
Every Block of Coal broken from a working place liberates a certain amount of Gas . . 122
How a fiery Colliery was cured of its fiery character 123
The Comparative merits of placing the Gutters for Drainage of Mines in the Coal and in the Bottom Rock discussed and illustrated 128
Driving the whole Gangway into the Bottom Rock and its advantages illustrated . . 130

CHAPTER XXII.

EXAMINATION OF BREASTS—MEASURING OF WORK—PARLEYS WITH THE MINERS.

	PAGE
Systematic Ventilation; Formation of the Stoppings of the Headings	134
Details of Ventilation; Tracing out the Air-current	135
Conversations with the Miners	136

CHAPTER XXIII.

GENERAL CONCLUSIONS, WITH A COMPARISON OF THE DIFFERENT SYSTEMS OF MINING.

The systems of "Long Wall" and "Board" and "Pillar" Mining described and compared	142
The men and boys who work the Mines, and the credit which is due to them	144
Ventilation, and its relation to the amount of Coal Mined in a given time; Practical details of Mining	145
Mines having a Dip of about 45°; Mining in Pennsylvania; Loss of Coal in the system of Mining	146

SECTION IV.

THE VENTILATING FAN—UNDERGROUND FIRES—ELLENWOOD COLLIERY, MAHANOY COAL BASIN, PENNA.

CHAPTER XXIV.

THE VENTILATING FAN—HOW IT SHOULD BE CONSTRUCTED AND ARRANGED—PRINCIPLES OF ITS ACTION DESCRIBED.

CHAPTER XXV.

UNDERGROUND FIRES AND METHODS OF EXTINGUISHING THEM.

CHAPTER XXVI.

A DESCRIPTION OF THE COAL VEINS WORKED AT ELLENWOOD COLLIERY, SITUATED IN THE SOUTH-EASTERN BRANCH OF THE MAHANOY COAL BASIN, GIVEN TO SHOW THE GREAT NATURAL RESOURCES OF THE ANTHRACITE COAL FIELDS.

INDEX 171

LIST OF PLATES.

	PAGE
I. Overman's Tracing, showing a plan of working by Board and Pillar, and Ventilating by Splits, and especially adapted for working out the Coal of Seams whose inclination is low, and whose thickness does not exceed six feet. (Frontispiece)	36
II. Plan of Board and Pillar Whole Coal Workings, Dipping. Scale 260 feet one inch	36
III. Ventilation of Coal Mines by splitting the main current of Air; Details which show the relative arrangement of the main Drifts, Stoppings, Regulators, and Crossings	39
IV. Our Deputy Overman	45
V. The Putter	47
VI. The Coal Miner "Nicking" his Jud and the Pony Putter	54
VII. The Miner's first step as the Trapper, with the Air-doors	63
VIII. The Miner's second step—the Driver; With the Pit horse and his gears, as used in the Mines of Northumberland and Durham, England	64
IX. The Overman	58
X. Mode of working the Broken by splitting the Pillars and duplicating the Juds	65
XI. Sketch showing the mode of Propping a Jud in broken workings	67
XII. Sketch showing the method of working the Juds off the end of a Pillar, and the mode of Propping up the Roof	67
XIII. Mode of working Juds off the Pillars to the rise when the Dip is considerable and less than 12°	71
XIV. An example in which the "Broken" District is worked simultaneously with a "Whole" District, which is followed at what is judged a safe distance	72
XV. Working Coal by crossing the Metal Ridges of the old Board-rooms — Isometrical Section and Plan	76
XVI. Sketches of Appliances used in the early transportation of Coal underground, with Ventilating Furnace, etc.	81
XVII. Interior of a Board-room or Chamber showing mode of Propping and the manner of Branching off the Road into it, and the mode of Propping a Slate or Shaly Roof	88
XVIII. View in the Gap or Ravine cutting through a mountain, with a Drift, the "Trip," etc.	90
XIX. Entrance to Gangway; View showing how a set of Gangway Timbers are made, how put up and Lagged, and the man who Digs the place for them	91
XX. Isometrical view showing general plan of Breast-rooms and Pillars, etc.	103
XXI. Sketch showing Starter's Battery, and Loading Platform, etc.	104
XXII. The Starter	105
XXIII. Sketch showing how Manways are built against the Pillars inside the Breast-rooms; How the Manway door is set and position of Monkey Gangway; How the Manway through the Stump is connected with the Manway in the Breast-rooms, and the style of Battery used in Breast-rooms, having two Schutes; and other details as per description	113
XXIV. Section of a Breast of Coal in the Mammoth Vein, together with Breast-room, Excavation, Crossheading through Pillar, Starter's Battery, running Schute, Platform, Break Stick, Schute Timbers and Wagon in position to be loaded	115

(xv)

THE ART OF MINING COAL

DESCRIBED AND ILLUSTRATED.

SECTION I.

GENERAL INFORMATION CONCERNING COAL MINES AND COAL MINERS, WITH DESCRIPTION OF LONG WALL AS WORKED IN HORIZONTAL SEAMS OF COAL IN ENGLAND AND IN FRANCE.

CHAPTER I.

EXPLOSION AT SPRINGWELL COLLIERY; MEANS TO BE USED AS SAFEGUARDS AGAINST ACCIDENT.

It is a painful sight to witness the effects of an explosion at a coal mine. While a child in small garments and of few years, I formed a tiny unimportant personage of a large and excited crowd which had collected at the mouth of the Springwell coal shaft. The Springwell Colliery belongs to the Liddle family, one of whose members, the late Lord Ravensworth, is often quoted in an honorable and laudable connection with George Stephenson, the Railway Engineer.

Above the shaft was a cloud of black dust or of smoke, whose thick unfolding volumes were spreading throughout the atmosphere to cast their dark gloomy shadows over the locality; and this ominous cloud had just been vomited violently out of the coal shaft!

What apparition more than this was necessary to cause that mining village in full view to become promptly deserted? A minute before the rays of bright sunlight had been flowing to the earth uninterruptedly; but the change came suddenly, and the contrast of the past and present was unnatural.

No one remained in the village who possessed the power of locomotion. In the grief-stricken crowd at the mouth of the shaft were a few men, many women and children—including babes clasped frantically in the arms of their mothers—who were among the first to reach the fatal spot. Many were the wails and sad were the exclamations they made! In the painful picture there was no background of consolation; and he would be a clever artist who could trace it truthfully with his pencil.

It was needless to tell them what had taken place. Their instincts told them that not a soul could live in the mine a hundred fathoms below while the shaft was brimming over with *stythe*—a term applied to after damp.

In the depths below were the sons of that venerable couple whose days of usefulness had been ended by the ripeness of old age. There were the husbands of those wives and the fathers of those children and the brothers of those younger women. And there, too, the bonnie bairn of the poor and lonely widow!

The alarming cry of "Springwell pit 's fired" spread over the surrounding country like a telegraphic wave, and this brought great throngs of people from the other villages in the district.

A child of four years could hardly understand much of what was transpiring; everything seemed to be of such a mysterious nature and so completely above his comprehension, as to bewilder him. The viewers came in their white flannel shirts and blue flannel clothes and drove the crowd back from the mouth of the shaft; and the banksmen got the pit ropes into their places, and, hooking up *curves*, called to the engine men to chase the ropes. So the curves and ropes were made to traverse the shaft in obedience to this order for the purpose of agitating the air in the mine and cause it to circulate and render it fit for respiration. No one could descend before this was done. Then a body of searchers stepped into one of the curves and clung to the few links of chain which were coupled to the flat hemp rope. When lowered beneath the "saddle boards" into the dark shaft, they were accompanied by the prayers of the crowd as the rope was paid slowly out of the slots in the engine house side; and crawling silently over the huge pit pulleys, glided slowly and steadily down the shaft, lowering its living freight to search for those dead bodies which were shrouded in the sulphurous atmosphere of the mine. But much had to be done before the air could be made to circulate so as to drive the black damp in advance of the current, and days seemed to pass over before the crowd of watchers at the shaft's mouth seemed to abate. By this time many a parcel had been drawn from the depths below and landed on the saddle boards, the coverings removed, and the contents, often burnt to a crisp, scrutinized and identified by some token, such as a portion of unconsumed clothing, or by some mark on a part of the body which had been screened from the action of the fire by the thick flannel clothing generally worn by the English miners. As for the exposed parts of the body, the face and hands particularly, they hardly bore a trace of human resemblance. But for all this, that widowed wife and childless mother knew the remains of her offspring as soon as he was brought to "bank" in the arms of a pitman. No power could separate her from the form of her poor bairn until the strength of her frail body succumbed to the superior force of her grief, and then she was borne off in the same direction as her poor boy.

This is what memory brings to me of a scene which happened between forty and fifty years ago, the closing features of which were the funerals which bore away the dead miners to their last resting-places. After the strange faces assisting at the funerals had vanished from the village, and a strange inexplicable influence was beginning to exert itself which affected young and old alike, and brought every one under its restraints, the dead bodies of the horses, cooked and charred, were drawn from the shaft and hauled away out of sight. This seemed to form the last of the picture, and it went away from my sight like a dissolving view; which, however, the mind can recall as it was then understood by a child, and as it has been remembered through his mature years, and it can be seen still as it appeared then, but modified by the shades that experience throws among its figures.

Brought up in a district where the recurrence of such accidents has been by far too frequent, I must confess to having yielded much to the impressions they have produced, which, in great measure, have influenced the course of a subsequent mental training. Much of the spirit those teachings have promoted will be reflected in the following pages, in spite of any effort I may use to curb and subdue it. Therefore, if my sympathies are shown to be largely with those, man and master alike, whose lot it is to earn their daily bread within the dark, dusty recesses of the mine, where Death sows his seeds broadcast, and reaps his harvest with a bloody sickle, the reader will please to consider in as kind a manner as possible those expressions and exclamations which have their origin more in the heart than in the head.

Since the occurrence of the Springwell "misfortune" alluded to above, very great have been the improvements introduced at coal mines, to be used as safeguards against accidents. In all well-regulated mines the officers know how important it is to attend to such rules and regulations as are those which follow.

Keep in advance all dead work, and where gas is very abundant, and the slips and crevices frequent, bore ahead holes not less than four inches in diameter, and twelve feet in length. By an examination of these bore holes daily, an idea of the quantity of gas contained in the coal may be formed, and sudden outbursts, to a certain extent, avoided. The gas will drain off through such bore-holes very rapidly, but with much greater regularity than if such bodies of gas were let out by the sudden removal of large quantities of coal, as in the case of blasting.

Divide the ventilating current as often as it is necessary to do so to keep the air in certain places sweet and respirable. Let the amount of coal cut in any particular district regulate the amount of air sent to it, more than the extent of such district. Thus, if you mine a hundred tons of coal in a *run* daily, send in the air to this run or breast in sufficient quantity to dilute the gas given off by the mining and breaking up of those hundred tons of coal. Let the air passing from such part of a mine get into a return air course in the most direct manner possible, by passing it direct to an upper level; or, if this is not practicable, pass it into a return air course, such as a monkey gangway, by carrying it over a main air course by means of an AIR CROSSING.

Examine the abandoned excavations, and note the state of the air within them. In old places you will find the air and the gas it contains so thoroughly incorporated the one with the other, as to form a dangerous mixture if the percentage of gas is over three at one time, and is greater than this at another. The gas in such places, when neglected, has done much mischief by strengthening so uniformly and so gradually in the air as to escape detection until an explosion of a gigantic order has been the ultimate result, which a little intelligent attention, and a small increase in the ventilating current would have warded off.

If you have spacious air ways, and these you must have in order to obtain good and sufficient ventilation, use regulating doors to govern the "splits," and lock them securely at each setting. Regulating doors should be at a point in the return where the air of its district is making its final exit before it joins with the main out-going current.

Use as few air doors as possible, and none at all in the main intake air courses which wagons have to traverse.

Take great pains with the air stoppings, and build them so well and bind them so securely that the force of an explosion will not carry them away.

Air crossings are better driven in the solid strata. When this is impracticable, they must be built in the most substantial manner. The arch is the usual form; but if they are not strongly bound and secured by cribbing, or by strong iron hoops, they are liable to be blown up by the force of an explosion acting under the archway.

Air courses should be as straight and direct as possible, and as capacious as they can be made.

Officers of mines should practise the art of discovering the presence of gas in the air and learn to judge of its proportions by the manner in which it burns over the flame of a candle. Two per cent. of carburetted hydrogen in the air can easily be detected by the flame of a closely-snuffed candle, or by the flame of an oil lamp being reduced to burn brightly at its minimum, which may be done by the clearing of the top of the wick and the pulling of it down so that its top will be even with the level of the tube. By the use of a flame of hydrogen, on account of its great heat, the gas in the air may be burnt and its flame seen if its proportions in the air are as low as one-half of one per cent. By burning this flame in pure air and then in a mixture of air and gas, the difference, which is marked, can very easily be distinguished.

If the ventilating force be acquired by any mechanical appliance, this should be duplicated, so that in case of an accident to one the other could be promptly used in its place.

All mines should be so well inspected daily, by its own staff of officers, as to render the inspection of them by any government officer unnecessary.

The safety lamps, hoisting ropes, timbering, and all other appliances should have their due share of attention; but defects in these appliances are to be seen by the naked eye and may be detected by any ordinary person; it is different with the air which so often floats the angel of Death into the presence of the unsuspecting miner, who goes to his doom, in many instances, sheerly because of his ignorance concerning the nature of this enemy.

CHAPTER II.

COAL-DUST AND COAL-GAS IN THE AIR FORMING A DANGEROUS MIXTURE.

We are told that the explosions of coal mines cannot be avoided. Sudden outbursts of gas break away through the coal and fill the galleries and excavations in their vicinities with gas, just in the same manner that the bursting of a steam boiler would fill the air in its immediate locality with steam. Large portions of coal are thrown out at the same time. Certainly no ventilating current howsoever large could overpower the large quantities of gas given out at such times; hence the necessity of driving all opening places well in advance of the others. Such outbursts of gas taking place in any locality where a large number of men are at work, enhance the danger of explosion.

An explosive atmosphere may be composed of various ingredients; but in coal mines, coal-dust and coal-gas form the active combustibles. Coal-dust, so finely pulverized disseminated through the air as to be almost imperceptible, forms a dangerous constituent and one which does not receive the attention it merits. It is a more unmanageable ingredient than the coal-gas itself. A swift current of air traversing a mining passage not having its sides moistened by artificial or natural means, gathers up the dust on its route and bears it through the working places.

In the thick beds of anthracite which lie on a high inclination and are worked by breasts, resembling a series of inverted quarries, large bodies of dust are formed by the coal as it works its way down to the gangway. But it is where the coal grinds itself down through the STARTER'S BATTERY that the great clouds of dust are met with in the air. At every rush of coal you see volumes of dust burst out of the battery and fill the passages with their opaque masses. Then, again, down at the gangway, where the coal is running into the wagon, you cannot see the flame inside of the loader's lamp ten paces away from him. Nay, I question whether there are not times when the loader himself cannot see the light of the lamp he holds in his hand.

A manway is a passage formed on the side of an excavation by a series of props being laid against it. The props are each secured, head and foot, in holes, one being cut in the rib about five feet above the bottom slate, the other in the bottom slate, about two and a half feet from the rib. At a distance apart of about five feet, they are laid against the rib so as to be as nearly as possible parallel to each other, and they are set at such an angle as to lie in the same plane. Planks nailed on the outside and spanning the distances between them form a space underneath, large enough for a man to crawl through. By being advanced with the breast and kept a few feet in arrear, it forms the travelling road of the miner; and it forms the air-course for the ventilation—one manway being on each side of the breast room. On the outside of the planking a portion of the mined coal of the working breast is stored until the breast is worked up to its limits; and this coal forms the gob of the coal mine. It contains elements of danger in the shape of gas, coal-dust, and occasionally the resulting gases of spontaneous combustion, among which that fatal gas, carbonic oxide, takes its place, with that other poisonous element, sulphurous acid; these uniting with others to form the WHITE DAMP so fatal to the physical constitution of the miner.

It is through this gob that the miner's travelling way is built, and it is through this that the air goes which ventilates his breast. If he have a large quantity of air through so small a passage as this, the current must travel at a rapid and consequently at a dangerous rate. The area of such a passage cannot exceed six feet, and when the miner gets inside of it or stows his materials and timber there, it becomes nearly choked. Allowing it to be the clear six feet, to get a few thousand feet of air in circulation per minute will require a rapid current and necessitate the employment of a powerful ventilator. Three thousand cubic feet per minute forced through so small a passage will require a velocity of current equal to eight and one-third feet per second. This is a dangerous rate, inasmuch as it will force an ignited body of inflammable gas through the gauze of a safety lamp and cause an explosion in the air on the outside of it. A current of air driven through the manways at this rate picks up pieces of coal as large as peas and pelts them into the eyes of the miner as he dresses up his ribs to extend his travelling way. At many places he must wear goggles to mitigate the punishment which the showers of coal-dust inflict. But what of his lungs? The strong men who work a few years in such showers of dust become weak, and their faces blanch from the continuous absorption of the coal-dust into their systems. A consequence is that you put your miner working in such mixtures under the sod twenty years prematurely. Saying this, we are putting the many other liabilities to which he is subjected entirely to one side. Those other dangers, which are evident to his senses, he may watch at will and guard against at discretion; but these, which are so subtile in their nature as to be far above his comprehension, or which do not receive a due share of his attention, are the ones to be held up in front of his mind's eye; and it is much the purpose of this work to hold up danger posts somewhat like those our deputies sometimes set at the end of the "fiery boards" to point out those dangers which are not so well and so generally understood.

CHAPTER III.

MARSH GAS, AN EXPERIMENT. THE EFFECT OF A BRATTICE IN AN OPENING SCHUTE.

In studying the qualities of matter, after completing our elementary course, we read to the greatest advantage from Nature's book. In other words, we get the substances themselves and examine for ourselves and thus become more thoroughly acquainted with the nature of them. The gases are the most delicate of all the ponderable substances to handle. They are the least manifested to our outward senses, and we must constantly advance on them well armed with the delicate instruments that science and art have placed in our hands. The coal mining engineers of Europe have been for a long time united in their researches after the nature of gases and in inventing means and appliances to be used in mining as safeguards. We have learned much from those researches. Mining casualties are attended not only by loss of life, but by great pecuniary loss as well. Consequently, it is much to the interest of a colliery proprietor to have established at his colliery all such systems as may enhance the safety of his mine. To show in what light miners regard gas, the relation of a couple of instances will not be out of place.

All who have studied elementary chemistry know something of the marsh gas. It is identical with the carburetted hydrogen of the coal mine. As we are not writing for chemists, there is no need to give a list of its chemical constituents. To know that to consume the gas as it comes from the marsh or fresh and unmixed from the pores of coal, there must be eight cubic feet of air to each cubic foot of the gas, is enough. To dilute the mixture so as to render it incombustible requires at least sixteen cubic feet of air to each cubic foot of the gas. To experiment with such delicate things you require a rather complete laboratory, and if you need any one to assist you, he should have some idea of what is going on in his presence. Speaking in the second person, with the kind reader's permission, will do just as well to relate those incidents above referred to, which are real in every particular.

You have a laboratory in a particular mining town which need not be named. Temporary as it is, it answers your purposes admirably. People do not know what you are about; and they say many things and draw many inferences, and these are always wide of the truth. People are not all charitable, and some do not know whether you are smuggling whiskey or coining false money. It does not matter how open you keep your doors or how well you entertain and amuse your neighbors by allowing them to read all your periodicals and by giving them all the honest information they ask for; you are not relieved from suspicion.

On your table stands a large jar of the marsh gas just now spoken of. To hold the gas securely, the jar is inverted and its mouth is immersed in water. This prevents contact with the air and preserves the gas from endosmotic action or diffusion. Patrick Cummings steps in. Pat has been a very worthy and honest acquaintance for some time past, and you and he have become quite familiar. His attention is fixed on the jar of gas.

"What the d—— is in the jar, shure?" asks Pat.

As full of mischief as Pat is of curiosity, you answer in a riddle which Pat does not care to solve.

"That thing in the bottle is our common enemy; the gentleman whose name you have just mentioned, or one of his imps at least."

Pat's attention is none the less fixed by this careless answer, as his face expands into an incredulous smile.

A wicked idea strikes you as you ask Patrick if he would like to see the gentleman in question.

There is no answer, but there is that expression in his eye which questions your ability. You accept it as a challenge and carry out your idea at you do not know what is to be the cost. You take a match from the safe and strike a light, and raising the jar hold the flame of the match to the mouth of it. That portion of gas in contact with the air ignites. Being pure, the gas in the jar cannot explode for lack of air. You turn the jar right side up, and then the gas, being so much lighter than air, rushes upwards, and a body of air, equal in volume to the escaping gas, descends into the jar to displace it and force it out. There is a slight struggle between the air and the gas, and then a volume of flame shoots up to the ceiling, carrying away with it the neck of the jar.

Patrick is so much taken by surprise, that in his excitement he invokes the aid of holy and sacred powers, in the middle of which you cannot refrain from profane and boisterous laughter.

You lose your jar, your gas, and your reputation at one foul blow of your own striking.

Patrick tells his tale, and he does not forget to comment on a point which adds nothing to your credit. That there is some mysterious connection between yourself and his satanic majesty, Patrick firmly believes; and he does not hesitate to impress this idea on the minds of many other persons who believe him.

Marsh gas in sufficient quantities for experimental purposes may be obtained in any place where vegetable matter is decomposing in a stagnant pool of water. All we need to do to get it is to provide a jar with a wide neck. Fill the jar with water, and while the mouth remains under the surface invert it. Then raise the jar, taking care to keep its mouth only below the surface of the water. Directly under the jar stir the mud in the pool with a stick, and secure the bubbles of gas as they rise through the water. The gas entering the neck of the jar will rise up and take a place over the water, which it will displace. As soon as the water is all out of the jar the latter is full of proto-carburetted hydrogen, and is sufficiently pure for ordinary experimental purposes.

In carrying the gas away from the pool of water a saucer should be placed under it. A portion of water in the saucer will prevent the air from coming into contact with the gas.

To describe the nature of the existing prejudices among miners and their bosses, we will relate the following incident.

There is a colliery within a short radius of Ashland mining its coal from the Mammoth vein. The dip of the coal seam is over 50°. Like many other seams of the Mahanoy basin, it is divided by numerous breaks. These divisions often run through the rocky strata, and have had their surfaces in contact ground smooth by their movement under great pressure. These form spaces into which the coal-gas has lodged itself, and as the coal is worked out, the gas in those spaces pours into the excavations in large volumes.

In the old abandoned places you have been robbing the pillars and have been earning high wages to the envy and disgust of the other miners. These have been wishing to have you placed in some work driven in the solid coal and thus placed on an equal footing with themselves. The time arrives when there are no more pillars to be robbed, and you find out the inside boss, who is not at all a friend of yours in his heart on account of your being on the most intimate terms with the "operator."

You, as an ordinary *employé*, are entitled to the first breast to be opened, and as the gangway has been driven well in advance of the last open breast, there is plenty of room to start off the schutes and manways necessary to open out a new breast of coal. The following statement will show the common mode of giving a contract to a working miner.

You apply for the working and opening of the new breast just alluded to, and the boss tells you that no *fan boys* are to be obtained; and until he gets some of these necessary assistants, you must wait an indefinite length of time before you can begin the work.

Fan boys were needed to turn the hand fans used to blow a current of air through square wooden boxes into the schutes and manways as these were driven up the incline, because the gas, being so much lighter than air, tends to settle in the spaces formed above the levels through which the currents of air pass.

You do not like the idea of losing your time for so trivial a cause, so you say to the boss in a tone which is entirely too frank for your own interests when a bargain is to be made, and you see your opponent taking a mean advantage of you, as he evidently was in this case, by making an attempt to get rid of you—

"Supposing that I can obtain fan boys, will you allow me the regular rate of wages you pay to them?"

This proposal could not be objected to, and the boss did not object to it.

But you make another which takes matters a step further; you propose to drive the fans yourself. This staggers the boss, and he hardly thinks you in earnest. You add another condition, to close the bargain, which is the only sharp thing you ever did in your life—

"In the event of dispensing with the fan boys, will you then allow me the amount of their wages?" you ask.

"Yes," says the boss, adding an expression which need not here be repeated, and thinking you to have lost your senses. He could not see how the gas was to be driven out if there were no fans to be used.

This settled, you find a *butty* and you start off the schute while your butty starts off the manway. After the first day's work you have a large space opened in each of the places. So the next morning you simply hang a piece of canvas across the gangway at the foot of each excavation. This has the effect of turning enough air into the elevated spaces to keep them clean. You work on, and as you advance the places, you join a brattice to the canvas; and this brattice carries the air into the face of each, and you have no trouble to keep them safe by the ventilation you thus obtain; and you find it to give you less labor than the fixing of air boxes would require. So your two dollars per day—the fan boys' wages—come to you as a perquisite.

What was thought of the arrangement by the miners and the fire boss, who lost his job of *brushing out the gas every morning* by it, the following will show.

You go to the mines late one morning, about nine o'clock in fact. When you arrive at the schutes you find the men collected there bent on having a row with you. The men are mostly Welsh, but a few Irish are among them. You receive a volley of abuse to begin with, and the *fire boss* joins the crowd against you, and you feel that you are getting into a scrape.

The miners who work piece-work in the same manner as yourself complain that your places have not been examined, and they will not work until they are. Then you turn to the fire boss and ask why the places have not been examined.

"Because, you see," says he, "you be working the gas out by contract, man dear!" and he places a space between every word of his sentence, and holds his safety lamp in your face to witness the effect his words—meant to ridicule you—have on your temper.

Now you are not the coolest man in the world, and between anger and disgust you can hardly contain yourself. For the sake of half a minute's work the fire boss has allowed a dozen men to remain idle for two full hours; and they are nursing a bitter wrath against you, whom he wished to hold up as the one responsible. You know that the fire boss is thought to be very sage, and is in a responsible position.

But you know also that there is a tolerable current of air passing, and that a large portion of it is travelling past the brattice and into the face of each of your excavations, and that there can exist in them no dangerous accumulations of gas. With this conviction you place your *naked lamp* on your hat; and before any one of that excited crowd can know what you are about, you spring on to the platform of the opening schute. No one attempts to hinder you; but being convinced that your places would be as full of gas as have been all other schutes after standing undisturbed over night, they scramble to their feet and vie with each other to get as far away from the spot as possible before the explosion that is to be takes place.

You crawl to the face of the schute and examine every nook in it; but you find no gas. The air circulates quietly past the brattice, and where such is the case, except very large quantities of gas are blowing out of the coal, no dangerous accumulations can take place.

You sit quietly down on the loose coals near the face, and your chagrin gives place to uncontrollable mirth. The footsteps of the crowd racing outwards so as to almost break their necks have scarcely ceased to sound in your ears, when it occurs to you, as the funniest part of the farce, that your butty and the fire boss have also disappeared and you remain all alone in your glory.

"Could all crowds be dispersed in this manner there would never be any riots and little need of policemen," you cannot help thinking.

But a step comes creepingly along the gangway, a light appears at the platform, and you discern a broad face turned up towards you. After a few seconds of contemplation you hear a rough voice call out, not to you, but to some of the others away out on the gangway.

"Bhoys, come here! By the powers, the gas is charmed!" and one creeps in after another, and the *monkey men*, the *gangway men*, the *crosshole men*, and the *schute men*, starting other schutes in an inside opening breast, *all* of whom have been kept idle by their distrust of your brattice, now slink into their work as if they were ashamed of themselves.

So you finish the schute and manways and headings without the aid of fan boys, and receive a bonus of fifty dollars for your services, and besides save the fire boss the trouble of brushing out the gas of a morning, which is a barbarous custom, and has cost many a life by sending the gas ready mixed on to some distant light, there to be exploded.

You would fancy that bratticing would have become the custom at this colliery from that moment, but it did not; why it is hard to guess.

But in a short time after this you urge Joseph Brown, then the inside boss at the Tunnel Colliery, Ashland, to try it in the same manner, and it works so well as to acquire a reputation which takes it into all the fiery mines in the coal region. But the brattice has been used in Europe since the time of Spedding's application of underground ventilation, nearly two centuries ago.

CHAPTER IV.

IMPROVEMENTS OF MINING IN THE SOUTHERN AND WELSH MINES. COAL MINING ENGINEERS. HETTON COLLIERY.

In the Lancashire district, the upper seams of coal gave out but little explosive gas, and the mines there were worked on a small scale. Many hoisting shafts were used at a colliery even in the present century, and the ventilation was not very perfect. You would see one engine in the centre of a group of shafts used to wind the coal from all of them. There was a system of drums and counter-shafts connected to the main shaft of the engine; and some of them were set at right angles to the main. The work had neither system nor centralization in it, and the yield of no particular mine had approached to one-half of that of the average collieries of the North.

After the sinking of some of the shafts through the rocks to a lower series of veins, a great deal of gas was encountered in the Rushy Park and Little Delph coal seams now extensively worked at the collieries near St. Helens and Wigan.

Those seams yield a coal of excellent quality, and they vary in thickness from three to five feet. But the men who had been employed as managers in the upper seams made but little progress in these below; and in spite of the means devised, the out-put of the collieries was limited.

Matters of business brought Mr. John Wales, well known as a mining engineer of the North, and Mr. Wilson, a wire-rope manufacturer of Haydock, Lancashire, together, and those were met at the house of Mr. Wilson by an opulent colliery owner having an extensive colliery near St. Helens. The conversation flowed naturally into the subject of coal mining. Mr. Johnson was the colliery owner, and he expressed astonishment at Mr. Wales's account of the Hetton collieries, whose admirable system of working and ventilating had been perfected under the joint efforts of himself and Nicholas Wood—the latter well known by his work on railways. Mr. Wood had been an apprentice of Mr. John Buddle, and also was the founder of the Northern Institute of Mining Engineers.

Mr. Wales paid a visit of inspection to the mines of Mr. Johnson, and he saw that no change could be effected until a complete change was made in the officials. The manager was on a par with the managers of our anthracite mines, a clerk and civil engineer, without a knowledge of mining in its details. The actual management of the mine was left to a man brought up in the mines, one of those *self-made men* we are constantly meeting with in all kinds of public works. They want none of your book learning—your geometry, and the language you use to comprehend it by, algebra; your technical science may go to the deuce for them; they do not require such nonsense stuffed into their stubborn heads. To get along with such men in a dangerous mine would simply be an impossibility. Mr. Wales, knowing this, invited Mr. Johnson to pay him a visit at the Hetton Collieries, to which invitation Mr. Johnson promptly responded.

At Hetton Mr. Johnson was astonished at the gigantic nature of the works, not so apparent on the surface as they were made manifest to him under ground, and at the plans in the office and at the explanations made by Mr. Wales and his staff of assistants and apprentices. But for these explanations Mr. Johnson would have thought that there were a number of unnecessary men used in the executive force.

First, there was Mr. Wales in charge of the collieries as *head viewer*. Then there was his assistant as general *under viewer*. Then Mr. Moore, the head mechanical engineer, in charge of the railways, engines, and repair shops, which had the magnitude of the repair shops of many railways.

In the mine Mr. Johnson found the coal of the rise to be lowered by self-acting planes, and that of the dip hauled out by engines winding it up the engine planes, and along the levels by horses drawing long trains of excellently constructed tubs or small wagons. He met the overman, whose great responsibilities were explained to him; and in fact he saw that no such system could ever be accomplished in mines except where the same division of trained labor was resorted to.

The systems of working out the coal did not impress him with so much force as did the manner of applying the forces. The ventilation of the mine, although an eye-witness, he could hardly conceive to be so great as it was represented by the scientific instruments he saw applied to weigh and to measure it. It was not so easy for him to believe that there were over seven tons (190,000 cubic feet) of air taken into the mine in the short space of each minute, and that this was maintained during the year in and the year out. That it was split up into a number of underground currents and sent into as many different districts without the aid of a door in the main roads, but through the aid of the regulator instead, showed him how admirable was the system of working and ventilation adopted at this colliery; such perfection being attained not in a day, but in a century, beginning under Mr. Buddle's direction, occupying the lifetime of Nicholas Wood, and now carried out through the careful superintendence of Mr. Wales. But the example of mining at Hetton was not in the north an exception as much as the rule; there were the large collieries of Ryhope, Pemberton, Thornly, and Haswell in the immediate neighborhood, and many others in the same coal field, the excellence of whose regulations was quite up to the standard of those practised at the Hetton Collieries.

From that day Mr. Johnson set his mind on the object of remodelling his collieries. Mr. Wales engaged, at Mr. Johnson's request, a northern mining engineer to take charge of his Lancashire collieries, and the engineer in turn provided himself with a staff of assistants. In a few years the improvements were so marked and the produce of the mines so much augmented, that other colliery owners in the district followed the example set by Mr. Johnson; and now in the Lancashire coal district there are many large collieries rivalling in extent and production the mines in the north of England.

But Lancashire was not the only district in the United Kingdom that received benefit from a change in system. Wales imported some of the northern engineers; and these brought along their assistants, and established themselves and their systems to the discomfiture of the Welsh bosses in the Welsh mines; and within the last twenty-five years the mining engineer—almost unheard-of in those mines before that time—has planted a branch of his mining institute, and, unlike the old boss, his predecessor, who kept all his paltry secrets connected with his craft to himself, disseminates that knowledge which is so much required by the miner, and which science teaches in the most liberal manner. The time may not be far distant when our bosses, who seem to be very much made up of those old-school bosses of

Europe, will be uprooted through the same agency and scattered in the same manner; I mean by the EDUCATED ENGINEER, let him be of what origin he may.

Being employed as one of a corps of assistants, I had an opportunity of witnessing the effect of the changes made at one of the Lancashire collieries. The mode of working out the coal as it had been practised we found to be admirably adapted to the district; and our books on mining call this the Lancashire Panel System. Mr. Hedley has written a valuable practical treatise on coal mining, in which he mentions it. In the ventilation of the mine we found the greatest defects. The machinery for hoisting, admirably adapted to its work, we found to be in its main points of AMERICAN origin and quite the equal, if not the superior in some respects, of its brother of the north made by the best of engine builders. I allude to the Stevens type of cam motion engine as used on our American rivers.

But the mine was divided into districts at once, and into each district a fresh current of air was sent to ventilate it. Deputies were employed to look after the working places and the proper ventilation of those districts; to timber the places, lay the track, keep up the brattice, and look after the work generally. Instead of throwing any more responsibility on the coal-cutter, he was relieved of that which he had; and his work was reduced to that of cutting coal only. He did not find any fault with an arrangement which relieved him of a burden and lessened his labor; he had no timber to hunt, and no rails to look after, no putter to pay. He lost no time on account of gaseous accumulations, and was not in a constant dread of his life. But he could not carry matches, nor light his pipe through the meshes of his safety lamp; and this was an arrangement he did not like, but to which he had to submit as a provision of general safety. In working back the pillars he was not allowed to leave stumps in the goaf to deflect pressure on the ends of the pillars as he cut them out. The coal was not to be crushed out by pressure and ground into dust; it was to be mined in lump in the best known manner; and the deputy was never so far away from the miners as not to see this provision properly carried into effect. On the whole, it was by the new arrangement of old plans more than by the introduction of new ones, and by the concerted action of the executive force that the increased safety of the mine was accomplished and the production tripled in a short space of time.

I was almost two years under majority at the time and considered at first in the light of an apprentice, but the requirements to be met in the organization of the mine were of such a nature and the executive force so limited, that a greater responsibility was put into my line of duties, and I thus became, at an early age, an acting assistant engineer. The working out of the new arrangements imparted many useful lessons.

In some of our books on coal mining the ventilation is proportioned to the area of the mine; that is, a certain quantity of air per acre of mine to be traversed is allowed. This may answer well enough where a large area of mine is standing on pillars, but it cannot apply to working faces from which varying quantities of coal are mined daily. We found in our practice that as we increased the production of coal in any particular district it became imperative to increase the ventilating current in a corresponding manner, and that it was necessary to remember that the miner picks out gas as well as coal with every blow he strikes. Although there was much gas in the mine we never made use of the safety lamp in any but the broken workings. Great attention was given to the state of the returning air currents, and the proportion of gas was never allowed to exceed a certain amount in them, and this amount being but two per cent. was so far from the point which forms an explosive and dangerous mixture as to cause no apprehension on the part of the executive force of the mine. The variations of gaseous discharges are so great that no rule can be made by which to work the currents in the different districts. The quantities of gas given off alone can determine this, and to determine the amount of gas given off in a coal mine or in any one of its districts requires the skill and judgment of an expert.

CHAPTER V.

THICK COAL SEAMS—WORKING OUT COAL IN FRANCE BY *REMBLAIS*—IN ENGLAND BY LONG WALL.

There is a rule among mining engineers which sets the yield of a coal seam at one thousand tons for each foot in thickness per acre, ample allowance being made for waste. At this rate the yield must be immense when the vein is upwards of thirty feet in thickness; and in many instances you find coal seams whose thicknesses are between twenty and thirty feet, and these are the thickness of some operated in Pennsylvania.

In Staffordshire, England, the coal seams of other districts have run into one thick seam; that is, the slates dividing them in other localities have in their centering in this locality gradually become so thin that the whole series of coal seams may be worked as one thick vein. Our own Mammoth coal is composed of several smaller seams, any one of which could be worked apart in a profitable manner were it to be found in a separate state. Sometimes this vein contains six *benches*, each having some particular characteristics.

In France there are coal veins of great thickness, and these dip at high degrees of inclination. After many experiments and much experience in actual mining, the French engineers have at last adopted that mode of working which they distinguish by the name of *remblais*. By its use they not only mine out the whole of the seam, but they are enabled to stow the refuse behind the working faces, and to preserve the surface intact at a less outlay than by any of the other methods they have used. This is accomplished by working over the seam, beginning at its out-crop horizontally and by packing the space behind the working faces with rubbish and stones, most of which are sent down from the surface, the balance being formed of the refuse of the mine. The working face then extends from the bottom to the top slates, and the floor is formed of the coal underneath. Props are set to secure what is overhead until the *remblais* has been laid, when all the timbers that can be taken safely, are drawn out. It is by setting props closely together that the *drawing* of timber can be performed with any degree of safety. The packing is done in the same manner as the building of the pillars in *long wall*; but the spaces are packed with greater care to make it as unyielding as possible when subjected to the great pressure; and instead of being limited in width as the gateway pillars of long wall are, they extend all over the excavation and fill it up entirely, excepting only the timbered roadways used to convey the coal to the surface. Then each succeeding lift has for its roof the *remblais* that has been built on the floor of its predecessor, and if attention is paid to the laying of the undermost layers, little trouble is encountered in the working of the lower faces; because on the large flat stones laid on the bottom, the matter above is so squeezed by pressure as to form an excellent roof, easily kept in position by a liberal use of props and planks and broad cappieces; and as the height is limited to two metres—about six feet six inches—the process of timbering is easily managed by two men working together. The greatest care must be taken at the extremities of the excavation, and the packing must be well rammed in to insure lateral support. Face after face is thus worked off as the work progresses downward into the coal vein, the light gas draining upwards as soon as liberated, and is carried off by ventilation before any accumulation can take place. Before this method was adopted the veins were divided off by a series of level galleries, each of which served to get out the pillar of coal lying above it. But it was found that the loss by falls of roof was so great as to cause this system to be abandoned and that of *remblais* to be instituted in its place.

In the thick coal of Staffordshire, work by long wall has been successfully instituted.

The coal is mined in *long wall* in a seam less than six feet by opening a face of several hundred feet in length. This is driven forward and the roof is kept up in the chamber formed by close propping. After the face has been advanced a certain distance and the weight of the roof begins to settle on the props, they are drawn out by the whole corps of deputies. It is not so much an object to save the timber in this instance as to get down the roof. As soon as the roof falls, roads are cut through the rocks, and pillars on each side are built up to that layer of top which has not fallen. These are approached by openings in the pillar of coal opposite to the working face, termed gateways, which are equidistant from each other, and the cuts through the fallen rocks open up a communication with the coal face and form a continuation of the gateway. After these gateways have been opened, the work of advancing the wall face begins at once, and cast iron props are substituted for the wooden ones. During the night a set of picked men go in and prepare the wall faces for the hewers and coal fillers of the succeeding day. These men build portions of the pillar up to the roof at each side of every gateway. They draw out the *metal* props the most remote from the coal face and reset them in a row parallel and close up to it. While drawing these props, if there is danger from a fall to be apprehended, a light chain is fastened to the props as they are knocked out, or in case of the roof creeping rapidly down, as it sometimes does, picked out by the digging away of the crushed stone on the head of the prop. The drawing is done by one man, while his mate attends to the jerking out of the props as soon as they are liberated. But these men know at what point the roof will break off when it falls, which it seldom does so suddenly as to jam them or maim them, although there are times when the roof falls over the props in spite of all their efforts to save them. But even then those props are not lost, because they are searched for by digging and are hauled out—when any of them are found—by the use of a lever and chains.

But it is in the gateways between the pillars where the gateway men often receive injuries of both a serious and a fatal nature.

As the roof crushes down it squeezes the pillars down to about one-fourth of their original height. In order to make height for the small wagons and ponies to and pass, a certain portion of the roof of the gateway must be cut away every night; after the walls have been prepared ready for the miners to work at on the following day, the gateway men employ the balance of what time they may have on hand before quitting time, in enlarging and securing these gateways. At some of the collieries the pillars are *waxed* to prevent fresh air from entering the grooves and thus promote spontaneous combustion, which, under great pressure, has in many mines set in and accomplished much mischief. The *wax wall*, or *sealed wall*, is formed by incorporating tough clay with the pillars as they are being built on the sides of the gateway. The clay becomes so thoroughly compressed as effectually to exclude the air of the ventilating current passing through the gateway.

As this mode of mining centralizes itself in a particular locality, from which several hundreds of tons may be mined daily, as many as fifty to one hundred thousand cubic feet of air per minute are required to keep the wall faces free from dangerous admixtures of explosive gas. The roof comes down generally, after the first fall, in immense slabs which lie flat on the floor of the coal seam, and the main breaks occur in lines nearly parallel with the wall face, which is always driven on the end of the cleavage in order to make the best WALL'S END or lump coal.

When the coal is mined in the Staffordshire thick seam by long wall, the top seam or bench is taken off first, and the roof being let down in large pieces, forms the roof of the other benches as they are successively worked off one after the other, until finally the roof and floor are brought together face to face, the operations in each bench being so nearly the same as to render further description unnecessary.

SECTION II.

AN EXAMPLE OF MINING OUT COAL BY WHAT IS TERMED THE BOARD AND PILLAR SYSTEM.

CHAPTER VI.

SHAFT THROUGH ENGLISH COAL MEASURES—WINDING IN SHAFT—ENGINE PLANE—HORSE ROADS.

LET us descend into the gaseous coal seam called locally the Hutton Seam. This seam of coal is one of the lowest worked in the Newcastle coal field. It is sometimes two and a half feet in thickness and at others four and a half. In this part of the seam it is about three feet six inches. We shall herewith describe one method of working this seam—that known by the name of BOARD and PILLAR.

The measures incline very slightly, the dip being four degrees to the east. The shaft is sunk vertically through the measures, which lie so evenly and are so compact that their upper layers of strata form, in many places, the bottom of a reservoir to retain the water and quicksand which lie at the bottom of the surface earths. In sinking a shaft through these surface earths, much difficulty is often encountered on account of the activity of these immense beds or *pools* of quicksand so thoroughly impregnated with water. As we descend the shaft we pass through a cylinder of cast-iron, well strengthened by ribs, and put up in segments and sections to suit the size of the shaft and the thickness of the quicksand bed. The lower part of this cylinder is well jointed to the solid stratum which supports the saturated quicksands. We descend through the sandstones (samples of which you see in some of the principal buildings in Newcastle-on-Tyne), and then the slates and rocks and coal seams; those seams of coal termed workable being the *Fire Quarter*, the *Main coal*, the *Maudlin*, the *Low-main*, and the *Hutton Seam*. The *Harvey* and other seams are further below. At the bottom we find the "onsetter" putting the *full tubs* into the cages, which have double decks and double tracks on each deck. The cages, therefore, carry four tubs—small wagons which hold about ten hundred weight of coal each, and the facilities for hoisting and changing the wagons are such as to allow four full wagons to be landed every minute during working hours.

Before starting away from the bottom, we shall wait till the descending cage lands on the scaffolding below the platform, and watch the onsetter bump out the empty wagons descending by striking them with the full ones he puts in their places. Of course you see that two wagons go into the cage at the same time, and that the *onsetter* manages one of them while his assistant, a fellow as strong and as muscular as himself, manages the other. When the cage strikes, two stout boys stationed on an opposite side loosen the catches—holding the empty wagons in the cage—just in time to allow the empty wagons to be driven out by the blow they receive from the full ones. Then the catches, which have held the empty wagons, spring up and secure the loaded ones in their proper position for ascending the shaft with safety. The operation has been performed with an expertness astonishing to those who do not habitually witness such operations; but quickly as it has been done no time was to spare, for the catches have hardly struck the wagons, before the lower platform of the cage is raised an inch or two above the platform and is dropped back on the *keeps*, which bring the tracks of the cage even with the plane of the platform on which other loaded wagons are ready to enter the cage in the same manner as their predecessors have done. Then, without as much as a signal, the cage starts off with its freight to the surface, where in thirty seconds it is landed on the keeps at *bank*, where an operation the reverse of what we have just seen is performed.

While the banksmen are "banking out" and dumping the wagons into the screens, and the onsetters are getting wagons placed in readiness for the descending cage (for we desire it to be understood generally that while one cage is ascending on one side of the shaft with loaded wagons another cage is descending on the other with the empty ones) we shall look around us before we go further into the mine.

We examine to ascertain the cause of a humming noise we hear a short distance off. We find it to proceed from a system of sheaves and rollers and pulleys which guide a wire rope on to the drum of a large winding engine. The engine is worked by compressed air, and its duty is to draw trains of forty wagons at a time up an inclined plane at the rate of fifteen miles per hour, more or less, as the demand on the services of the engine requires. In a minute the train arrives and is landed in the full track of a siding, and the rope is unhitched automatically. By its momentum the train runs into a by-track, having a slight grade towards the platform of the shaft, and there it comes to rest. From this track the *onsetter* receives his supplies of loaded wagons. In the mean time the rope of the winding engine has been hitched to a train of empty cars, the drum of the engine has been thrown out of gear, and by gravity the train is descending the inclined plane (engine bank) of the mine; the motion being regulated by a brake attached to the drum of the winding engine.

We find this inclined plane or engine bank to be over one-third of a mile in length, and still advancing towards the boundary "on the dip" of the mine property. At the bottom of the engine plane a siding is made sufficiently long to collect into *engine sets*, the wagons brought out of the districts by the horses and the larger ponies.

The height of the engine plane is but four feet, and the height in the siding and horse roads is not less than five feet six inches. This latter amount of height is made by cutting up the bottom rock.

In order to get a horse into the levels, he is tripped over on to a truck and lashed securely on his side, when he is run as a passenger down the engine plane, and released at the bottom of it, where there is height for him to stand up. He is then set to work in the levels to draw the coal out from the districts, and his home is in stalls cut in the coal strata near the foot of the engine plane.

The mine ponies, some of which are so small as to be able to work in a height of three feet, have their stables near the bottom of the shaft, and they scamper off to them after their work is finished, running pell mell through the low passage of the engine plane, their drivers following in a stooping position.

CHAPTER VII.

DISTRICT AND PANEL WORKINGS—BOARDS AND ENDS OF COAL—WORKING LEVELS— DISTRICT DETAILS.*

In the mine we have several districts formed by main drifts dividing the coal up into as many panels. Each panel is worked independently; that is, it has its own set of hands, and is ventilated by an independent current of air.

* Plate I. is an overman's tracing, and it shows how simply the excavated parts of a mine may be represented by mere lines. Such tracings are the working plans, and you will often find them in the overman's pocket or in his cabin. When they become obliterated or worn out from constant use, they are easily replaced. The arrows show the direction of the ventilation. The doors, the stoppings, and the regulators are also represented. The use of the tracing is more for the mine than the office. Sections of mines are often thus shown.

Plate II. is more descriptive; and, although it shows all its excavations to be advancing and to be on the dip side, it very nearly resembles Plate I. in its details. It shows the manner of ventilation. It does not show any connection with the *rise side workings*. In a deep mine worked on this plan, the shafts are always carefully located in regard to several things. If the shafts are nearer to the boundaries on one side of the property than on another, it is often because that as soon as those boundaries are reached the *broken* can be commenced, and a larger yield of coal obtained, if such is desirable. *The air stoppings*, shown by broad lines drawn across the ends of the places through which no air must escape, show how the air inwardly bound is confined within its proper channels. *The air crossings*, whose use is to pass the foul air returning from the workings over the main passages which are conducting the fresh air inwardly, are shown in their proper positions near the entrances of each district. It is difficult to conceive how air could be *split* extensively in any mine without such a contrivance as the air crossing. The main crossing, throwing all the air coming from the south side of the main roads, is generally situated near the shafts and at a point where all the air coming from the rise and dip unites and passes over the main road to the upcast air shaft. When pillars are left of great width in the broken, the work is begun by *splitting the pillars*. If the dip is moderate, a *double turn* is laid at the end of the pillar to be operated on, and *juds* are turned off to the right and left. The work of taking out pillars is thus concentrated, and all the coal mined is transported through the *splitting* drift within the pillar. To understand how this is done, we would refer to the Plates X. to XIV. inclusive. Plate XI. shows to the left the edges of the strata above the coal seam, from which the roof of the juds, already worked, have broken and fallen. You see the props under the roof with their *cap pieces* intervening. Plate XII. represents the two juds and the two branches of tramway running into them. By removing a portion of the roof, we see how thickly the props are set to keep the roof up until the juds are completely worked off.

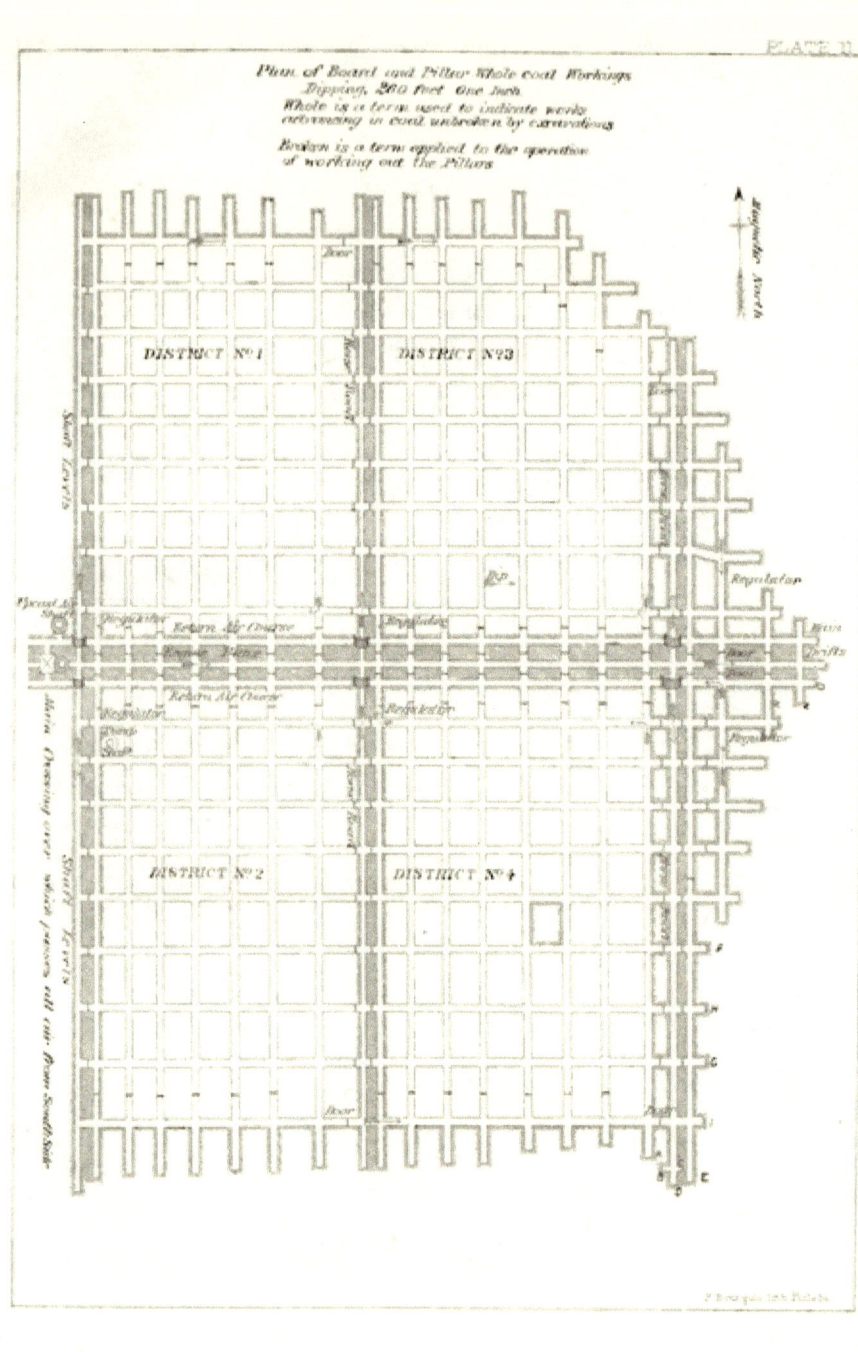

DISTRICT AND PANEL WORKINGS.

The plan of working adopted is known by the name of *board* and *pillar*, or *board* and *wall*, one of the most popular methods of mining we know of, and one which the working miners always prefer. It has its advantages, and in new coal fields where skilled miners are not abundant it is always adopted, although in England it is fast giving place to *long wall*.

The cleavage of the coal runs in parallel lines almost with the magnetic north, and its facings are so regular and so well defined, that the flat pieces they form, when split up, are so like boards of wood that the term board has been used in connection with them. Hence, any mining excavation driven across the facings is said to be going boardwise, while those driven at right angles to them and on the end of the cleavage, are going endwise or headwise. The terms have been corrupted by use, and we hear the terms *boardways*, *headways*, and *endways* to take their places in the mine. In blocking out pillars, however, particular attention is, or should be, paid to the lines of cleavage; two sides of a pillar are parallel to the main facings of cleavage, while the other two are parallel to each other, and most frequently at right angles to the facings of cleavage. The dip being from the west to the east, the line of cleavage north and south indicates that to block out the pillars, excavations are driven level endwise, which are intersected at certain distances by others driven up or down the incline of the seam and across the facings. The distances apart of these excavations form the limits and boundaries of the pillars. For fuller information we refer to the descriptive plans illustrating board and pillar workings.

Plates I. and II. are ideal plans of board and pillar workings made here for the purpose of illustration. We have the hoisting shaft in line with the main wagon road drift. The up-cast air shaft and the pump shaft are connected with the air courses and water levels in the manner shown. All the working places excepting two panels to the south in the *overman's tracing* are shown to be advancing, and the mine is shown as one mainly being opened.

After the shafts have been sunk, the main drifts have been struck off and advanced, as shown by Plate II., to the dip. Other drifts have also been driven to the *rise* in a corresponding manner. We have not shown the rise workings in this plan. Then a great advantage in using this mode of working coal mines is gained by the facility with which a large number of working places may be opened within a short time after the shafts have been sunk. By working mines according to our ideal plan of the dip side workings alone, we could employ one hundred miners in the "whole" coal, and these should mine from four to five hundred tons of coal per day. Also a panel or two of broken coal could be worked to double the out-put. If the same number of places were opened on the *rise* and the same number of men were to be employed in them, the out-put would be about eight or ten hundred tons of coal per day from the "whole" workings; and the product would be always on the increase until the required amount of business was done.

Our plan shows the boards to be driven to the east. This would not be a good practice to follow if the inclination were more than one in ten; because the greater portion of the coal would be mined from boards driven down the dip, from which the coal would require to be drawn up the incline. Yet the Shetland pony gets along very well in hauling coal from places having a dip of 1 in 8. To avoid this, it would be necessary to push the horse roads forward and start the boards off these and drive them to the rise, using brattice to ventilate the working face of each. After this general reference to the Plates I. and II., let us go on our way through the mine and note down what we see and hear.

We have said that at the bottom of the engine plane is a long siding which is used to collect the wagons to form the "set" for the engine. It is not at the bottom of the plane alone that a siding is provided in a mine that is more extensively opened, but in each of the horse drifts leading to the districts we find one provided for a similar purpose, and by means of switches the empty "set" of the engine plane may be exchanged for the loaded ones collected in the different levels. As the districts in the "whole" are similarly worked, a visit to one of them, to give a minute description of it, will serve our purpose and be a description of the rest also.

Starting from the siding in the engine bank, we turn to the horse road leading to the south, and although it takes us to a newly opened district, we can well understand by a reference to it, the plan of working the *whole* coal by the once popular board and pillar system. And we have here a very small number of working places, only eight, counting those marked at *A* and *B* in Plate II. which properly belong to the district west of it, but from which the coal mined in them may be taken to the horse road by the holing near the face of the headways in which the *air door* is placed. After the main headway D has been connected to A by B, a permanent *stopping* is built in the place of the air door which is transferred to the newly *holed* board B. But we have got so far ahead by entering the district thus, that we have forgotten to point out the air crossing near the main drift. Here we simply find ourselves under a common brick arch well cemented and made air-tight at the point where it joins the wall sides of the drifts at each end of the arch. Plates II. and III. show the form of air crossing used in the English mines.* In this country a crossing may very readily be constructed of timber in a more substantial manner than when of brick. However we are under this crossing, and we see that an air current is entering the horse road and passing through under the arch. If we trace this current of air we find it to pass into the face of the horse road, and by means of brattice and the innermost holing, it is carried into the faces of E, F, G, H, and I, and then it passes, after ventilating four more boards and one headway, as shown over the top of the arch or air crossing. This is the air crossing which takes such a prominent part in the ventilation of coal mines. As we pass along the horse road we see the drifts stopped off from each other at all points of intersection of the boards with the horse roads by the air stoppings, whose positions are indicated, as samples of all the other stoppings, by the lines drawn across the passages. The stoppings on the west side of the road keep the two currents of air, ventilating these two adjacent districts, apart from each other, and these currents may be traced to their destinations in the *return* by the arrows pointing out the direction of the currents. The stoppings to the east of the horse road serve to force the current of this district into the innermost holing, which is left open to allow it to pass into the working places of the district. It is well understood by those versed in practical mining, that when the headways D and E are sufficiently advanced, they are connected by another cross-hole similar to that now seen to be open through which the current is passing, and through which the largest portion of the current would pass in the event of a new holing ahead, were

---

* Plate III. shows, on a large scale, the road connections to the main tracks, and how the ventilation is effected by means of the air crossing. In some cases air crossings are made of boiler plate, and in others the arches are packed on the outside to prevent them from being blown up.

a stopping not built in it to force the air forward to the newly opened connecting passage. After leaving the working places of the district, the current passes a REGULATOR, which may be of any contrivance (a door for instance) that will allow the contraction or enlargement of the passage to be effected in the most simple manner. A door set up like a butterfly throttle-valve would answer well as a regulator.

We are now treating the air current as we would a fellow-passenger who accompanies us on a journey and with whom we wish to exchange our attentions; and we cannot say anything in this place scientifically of the intricate subject of ventilation. We may thus become acquainted with the mode of working the air currents in mines without the aid of that algebraical formula, against which some people are so strongly prejudiced.

Before going further, we must say a word of the Regulator. It is a throttle-valve by which the air of the district in which it is placed is regulated, and it may be opened or closed to cause an increase or a decrease in the quantity of air flowing to the district in which it is placed. But when an increase is effected it may be observed that this increase is obtained at the expense of the other districts, taking for granted that no corresponding increase of the main current descending the downcast shaft is effected at the same time. However, this is often a necessary step to be taken when it is found that the gas generating in some district is on the increase, and more than it is in another; hence the necessity for the use of the Regulator, which is an institution that has *grown* into mine ventilation almost entirely within the present century. And of its uses few people, even those who profess to understand the subject of mine ventilation, seem to have more than a superficial idea.

"The Regulator in ventilation," says one who knows, "can be only understood appreciatively by one who obtains a close practical acquaintance with it. A violin can be better understood by its player than by a mere musical theorist. It is thus with the Regulator in ventilation. It is best understood by those who operate it. Well, here we have the Regulator placed at the points indicated on Plates I. and II., and when we find too much gas in the current *at* those *points*, we use *our judgment* and open the regulator to allow the passage of more air; and on our next visit if the gas is not diluted to a safe degree we open it a little more; and so on until we are satisfied with the result."

How we ascertain what is the quantity of gas in the air we would like well to tell our readers; but much practice in this is required to become an expert (and none but experts on so important a point should be allowed to apply so delicate a test), because it is most accurately done by the *naked flame* of a candle, or a safety lamp with the top unscrewed. But such tests may safely be made by almost every *deputy* overman who is brought up in the Newcastle coal field. He acquires the necessary skill by his *everyday practice* and from the lessons he receives from his elders and superiors.

After we call your attention to the names of the places in the district, we may chat with the deputy in charge of it and take down some of the information he gives us.

The places D and E, Plate II., are the main headways of the district, and they are driven in pairs. N, just starting off from one of the exploring drifts O, is a heading or *wall* which is to intersect successively the boards I, H, G, and F, and those others intervening also which have been set off and are driven from the main headways. The intersections are made for the purposes of blocking out pillars and for ventilation; and where the gas is abundant the intersections are more frequently made, and the pillars are consequently of a correspondingly less area, which is not always a feature to be desired, as it exposes the sides of the pillars to the action of the air and reduces their strength in the event of a squeeze or crush, whose action is locally known by the name of creep.

We are again in at the district siding in which the wagons from the miners are collected in sets. The wagons are brought out by hand-putters in this case, and they carry ten hundred weight each. There are only two putters at this district, and we will see how the deputy places their work.

CHAPTER VIII.

A DEPUTY'S EXPERIENCE—DETAILS IN WORKING AND VENTILATING A DISTRICT.

The deputy of the district is a muscular fellow with an intelligent countenance. He has a habit of addressing every one in a suit of blue flannels as master, and blue flannels are the clothes one wears in the entire coal field of Newcastle-on-Tyne if his position is above that of deputy. The deputy's suit is of stiff white flannel invariably. His cap is of strong leather, and in shape not unlike that of a horse jockey.

"How are you getting on, lads?" we ask.

"Oh, I'm nicely, maister, thank ye! and, considering that we are just opening out work, we are doing finely."

"How much coal are you sending out?"

"Twelve score a day, more and less," he answers.

"How many working places produce this amount of coal, which, if I am right, is one hundred and twenty tons a day?"

"I have eight working places—three headways and five boards all in the whole coal—in which sixteen men work, following each other in two shifts; that is, eight men come in at 3 o'clock A. M., and the other eight follow at 10 A. M."

"You are in the foreshift. What time in the morning do you arrive?"

"About 2.30 A. M. I have time to get around the places before the miners come in."

"Do you find any accumulation of gas of a morning?"

"Never a stagnant accumulation; but always gas in admixture with the air. The gas in the air varies much, and requires close watching."

"What are the reasons for this variation?"

"Sometimes a fall of the barometer, denoting a reduction of atmospheric pressure, is the cause, and at others a change in the condition of the coal seam. If the coal gets a little soft or faulty in any place, the yield of gas increases in that place. It may be that two or more of the places strike such coal in the same day; then the indications are very marked, and the Regulator must be resorted to and shifted in a degree to correspond to the increase of air required to keep the district safe, and the gas sufficiently diluted."

"How do you know that, while you are shifting the Regulator in such an event to get more air into this district, you are not reducing the quantity in the other districts to so great an extent as to render them dangerous?"

"I can easily ascertain this by making an examination of the currents coming through the Regulators from the other districts. But in the case of a sudden outburst of gas, which is of rare occurrence, except when we plunge suddenly into faulty coal, we have plenty of time to consult with the deputies of the other districts and with the overman before we need shift the Regulator at all. By a daily report, we are enabled to compare notes, and we obtain, by doing so, a good idea of the condition of those other districts. We may know, if we like, what amount of gas is given off in each district, as well as what amount of air is in circulation in it. By averaging the percentage of gas in the various air currents, by summing them up, we may find out the total amount of gas given off in any coal mine where the ventilation of it has been reduced to a regular system."

"You know at all times what the total amount of air in circulation is in the whole mine?"

"We have a tell-tale anemometer, near the main intake, which gives the strength and velocity of the air current. We know the area of the passage at this point, and at a glance each morning, as we pass, we are advised of any serious variation in the amount of air in circulation. The variations are very slight indeed at this mine, which is ventilated by the new fan of forty-eight feet in diameter."

"How do you measure the air in your own district? Is it by one of those revolving and recording anemometers?"

"No, thank you, maister" (our deputy smiles sarcastically); "those instruments are for apprentices and other inexperienced people who want to find out the amount of air in circulation, for some purpose which is not exactly practical. But when we want to know at all times and in any place the condition of our air current, we must have some readier method of getting the information than that afforded by one of those delicate anemometers. We want to have a method of determining such an important point at a glance. An engineer will glance at his steam gauge at those times when the working of his engines slacks off. If his steam is all right, his engine is at fault, and demands his attention. If a serious reduction in the velocity of an air current were to take place, there are some men who declare they can *feel* that such a thing has occurred. It is a sensation which grows on a man, whose chief daily duties are very much concentrated in the care of an air current, if we may apply such a term. I have heard a marine engineer, in charge of an ocean steamer, say that he could tell to a pound per inch what pressure were in his boilers if he had a small leak in his steam connections, or if he could hear the water or steam escape at the opening of the boiler-gauge cocks. I think it is the same with myself and mates in regard to an air current to whose action we are daily exposed, and whose motions we are constantly watching, in order to detect any accidental variations of it. I fancy it is a sensation which gives us a notice to try the air when we seem to have half forgotten this important part of our duty. I have on two or three occasions advanced from a point where the full current of air was working to another where the current was slack, and have not gone more than a few steps— even with a preoccupation of mind and when in a hurry to get after some other business — before I have detected the fact of something being amiss. On one occasion of this kind, on examination I found the cause to be from an injured stopping, on another an injured brattice. Of course the leaks in each were very considerable, and the gas had accumulated in certain places sufficiently to render the air in them explosive."

"Then you object to the use of the revolving anemometer, and trust much to your sensations and experience?"

"Not altogether; good instruments may be used where very nice measurements are required. But then at least two instruments should be used, the one to correct the imperfections of the other; and, in case of their disagreement, a third instrument should be used. It is very likely that, for everyday use, our old reliable tell-tale anemometer will keep its old place at some prominent corner where its action can be seen by all passers concerned in the safety of the coal mine. You see this one, which I keep stationed here, gives me all the information I require of the ingoing current of air. (See Plate IV.) Its vibrations

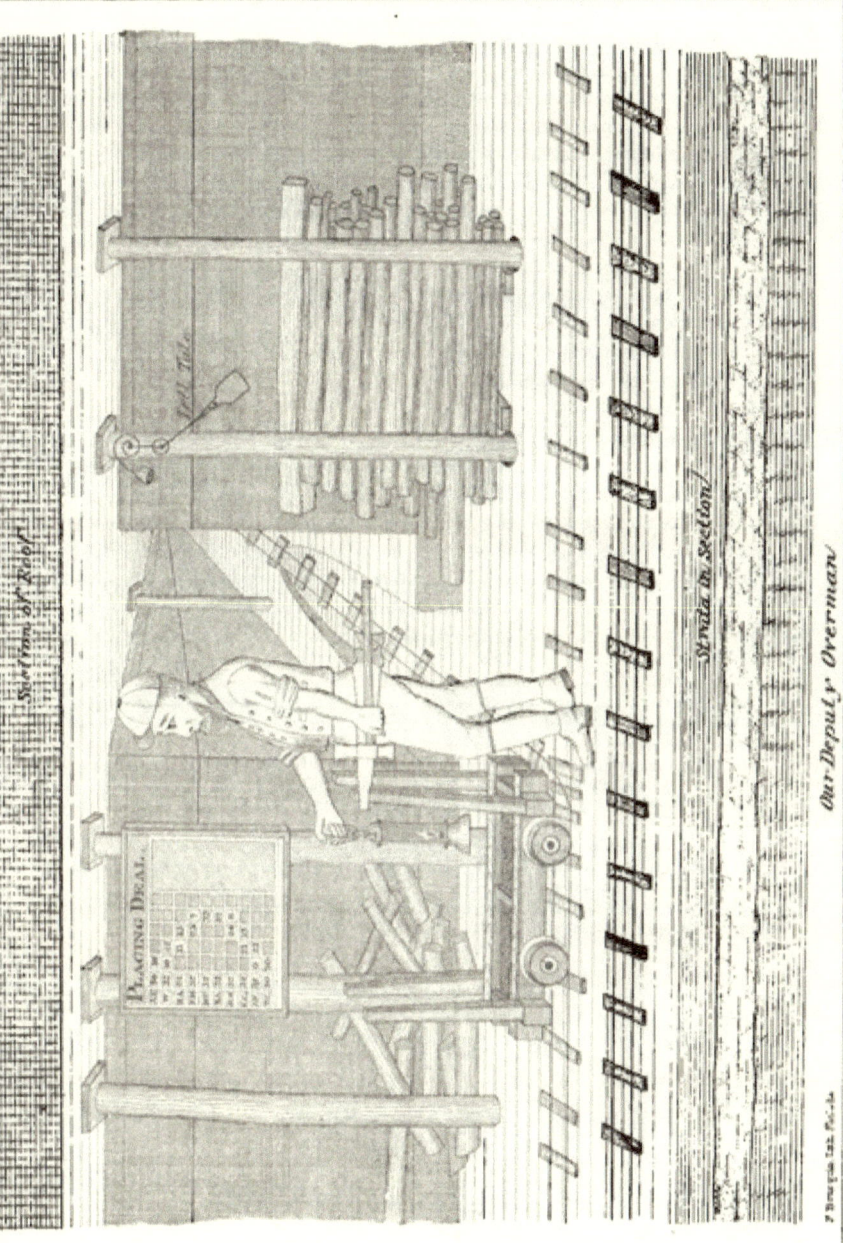

are produced by the motion of the engine trains and sets of wagons moving in remote parts of the mine; but there is a point at which it inclines always to be stationary when not influenced by the partial blocking up of the passages by the wagons and by an accidental blockade caused in the main roads intercepting the air currents, either through falls of roof or blockades formed through a train of wagons getting off the track. At such times, which are denoted by the irregular motions and varying positions of the *tell-tale*, the closest attention is required in those districts where we have copious discharges of gas; and at such dangerous interruptions of the current this simple *tell-tale* gives due notice, being attached to an alarm which jingles off and demands immediate attention. The figure in Plate IV. shows this simple form of *anemometer*."[*]

"Supposing a reduction of the air current were to be indicated here at the present time, how would you know whether the cause was by a blocking up of the air courses inside of us or outside of us?"

"Some people can tell by the whistling of air through a small leak in an air stopping. In case of a blockage occurring inside of any particular stopping, the whistling of the air through the leak is augmented. We soon get used to the music of any particular leak, and by a proper attention to it we may be advised of the locality of any serious blockade in the air course. This fact has led to the use of the water-gauge, which is a bent glass tube, having a contracted passage at the bend to prevent oscillation. When this instrument is applied to a stopping so that one of its tubes is fitted to a pipe passing to the opposite side of the stopping,

---

[*] Plate IV. The deputy, of whom much mention has necessarily been made in the text. Our sketch shows him as he comes out to the station each morning after he has left the men at work and paid attention to some of the most dangerous and other places requiring his services. He now comes out with his *horney tram*, which he is to throw off the track out of the road until it is further needed. He will place the lads' work and wait until he sees them start. This they do at once and with a race as soon as they come in and strip and have their work read off to them after their cavils have been drawn. We present the deputy with his vest and overshirt on, which he wears in spite of the heat and dust which cause him to perspire so much. But then the vest, which contains two capacious pockets, is an indispensable article of clothing, and is as necessary to our deputy as the tools he carries, inasmuch as these strong pockets are crammed with the "plate" nails he uses or leaves in the miners' places, samples of which you see in the bottom of the horney tram! The deputies are picked from the best men of a colliery. They are good workmen in all cases. In some cases they become students of the art of mining, and from their ranks are furnished very generally the chief overmen.

the water rises in one of the legs. The difference in level marks the resistance encountered by the air current while travelling in from a point on the outside of the stopping to the workings, and then returning to the point on the opposite side of the air stopping at which the instrument is applied. A variation in this gauge shows very often the way to an air blockade."

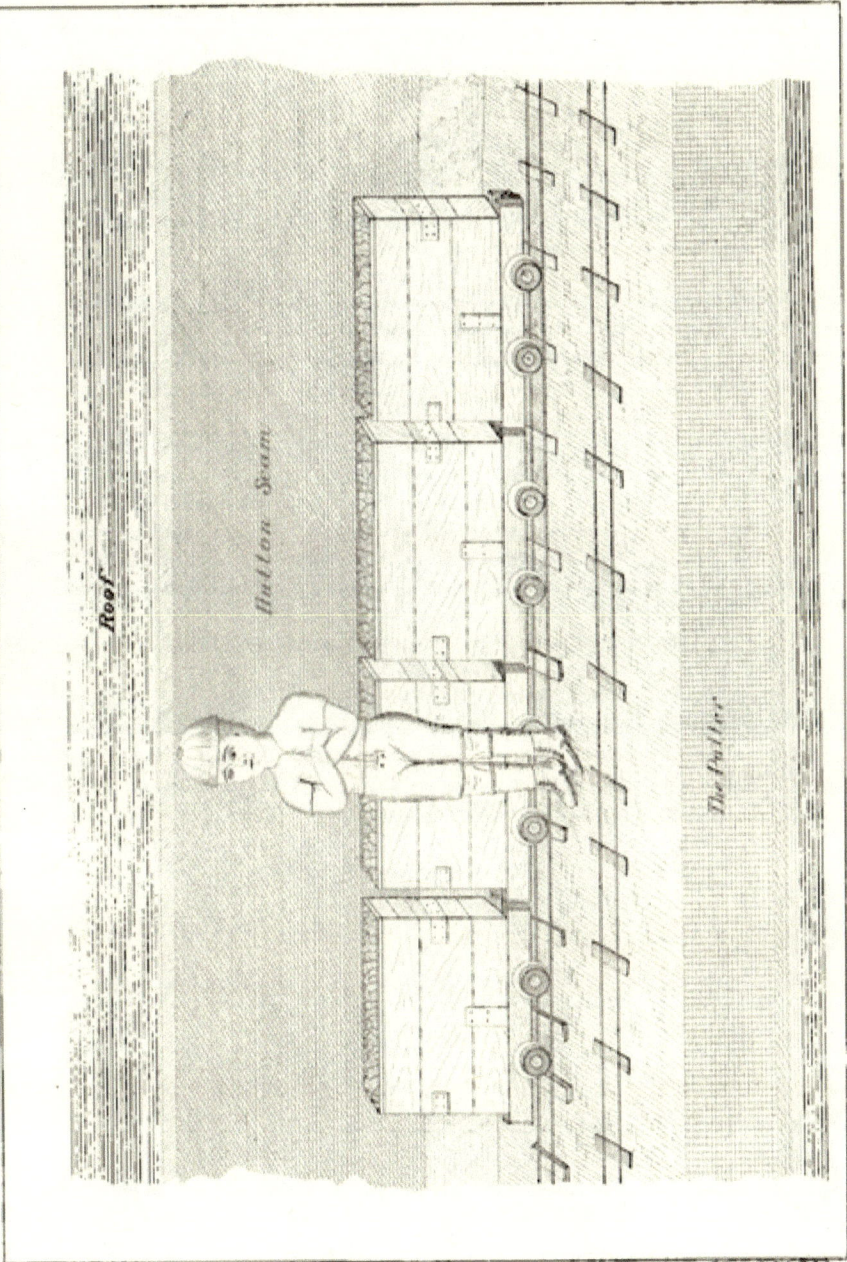

CHAPTER IX.

THE PUTTER.

"But you have other duties of an important nature to attend to besides those in which the ventilation of the mine is concerned?"

"Certainly; we have the roof of the roads to attend to and examine, and props to set at the face of the boards, and the road to lay after the mining faces, and the tracks to keep in good repair to keep these wild putters in their proper senses."

"Ah! I see you have a whole list of information to give me! Let us classify it, please, and begin first with this interesting feature, *the putter*, who, in spite of his light cotton cap, his sleeveless *body shirt* and *little breeches*, strong shoes, hoggers and *shoe* clouts, is in a lather of sweat; he cannot be passed by with merely a passing notice." (For sketch of putter see Plate V.)

"Oh, yes, maister," sighs the deputy, "these lads are very trying at times with their unreasonable demands on my time and attention. But their task is a severe one, and they have good, willing hearts, rendering their services in the most generous manner. We can forgive them all their saucy reproaches and mischievous tricks. Their muscles are being hardened for the work of the hewer of coal, which is even more severe than the work they now perform, which is fast becoming the work of the little *Shetland pony*, who with his driver is at the best only equal to one of these muscular half-naked lads. I had a lad, maister, lost at an explosion of gas at the Haswell Colliery, and I cannot see my putters here without thinking of the putter I have lost myself. He was a promising lad! The 'Call for the deputy; there's a plate loose,' only stimulates me to perform a simple, but not an unimportant part of my duty. It is certainly aggravating to a putter when his wagon runs off the way at a bad joint or a loose plate. It requires a pair of good strong arms to lift one of these tubs carrying ten hundred weight each on to the road again. Yes, I'll tell you of the putter, maister.

"After I have set the men to their work, and have given attention to the most dangerous places of a morning, the lads come in clamoring for the work to be placed, and the *carils* to be drawn. Here I have only two putters; but I shall speak of them as we have them in a large district, where there are half a dozen of them. You see this blackboard on which we put the miners' name in initials, or *numbers* in a vertical column, in the order of their remoteness from the *district*. We put opposite to each miner's name the number of wagons we allow to be mined in his place per day. Thirty-two for the boards, and twenty-four for the walls. These numbers form a column of figures which we sum up and divide by the number of putters we have. Here we have four walls and five boards, and the sum reaches 256. This, divided by 2, gives our two putters a day's work of six score and four each, which to get out, keeps them almost constantly on their legs. Now, as some of the places are much nearer than others, we place the putters' work in the following table. It will be seen that the third column gives the first putter all the work mined in the two nearest places, with eight in the third

Miners' places in the order as they are remote from the district siding.		Number of wagons of coal to be cut in each place.	First putter's work.	Second putter's work.
1	I	32	32	
2	C	24	24	
3	D	24	8	16
4	B	24	..	24
5	E	24	..	24
6	A	32	..	32
7	G	32	..	32
8	H	32	32	
9	F	32	32	
		2)256		
		128		

nearest place. It shows, also, that he gets all the work of the two farthest off boards. Thus, he brings the coal from the nearest and farthest off miners to the extent of one-half of his work in each case. The second putter takes the coal of the other places which are closer together, and are neither the farthest away, nor the nighest at hand. In this manner the work may be fairly placed, if there were a dozen of putters in the district. The letters have reference to those of Plate II., and they represent the places constituting the district.

"Have you trouble occasionally by 'using' this tabulated form of 'placing' or of arranging the work of a district?"

"Little with the older boys, who read off their work as soon as it is placed, and as soon as they remember what are the amounts of tubs they get at the different places, they start off at once with their wagons, the first putter taking the first empty wagon in to some of his men. At this district they are allowed spare wagons, technically called 'led tubs,' which each putter runs into his men as quickly as he can, taking only one at a time. When these 'led tubs' are all taken into the faces of the working places, the daily routine of work begins. Each lad takes his empty wagon into the turn-off, at the end of the board or wall he goes to, and brings out the spare wagon which the miner has filled. A strong putter, well skilled in the handling of his tubs, will come out of the board with a run, and on coming around the curve of the turn-off, whose radius is (for this gauge of twenty-one inches) for the short curve only three feet, he grasps the empty wagon with one hand, and pulls it back to the points; and often, without stopping his full tub, will start the empty one into the board with a run, singing out for the hewer to 'catch her,' while he runs after his loaded tub, still in motion, and starts off with all his might along the headway's course, wheeling around the curve and through the stenton, and from the stenton into the *mother-gate*, or main headway's course; and through this into the district siding, joining up against the train collecting into a horse's set. He then takes the first empty wagon in the empty track of the siding, and puts his tally or token on a staple driven in one of the corners inside; and he starts off again to *put* another tub. To avoid collision with another putter coming out he 'calls' at every few steps, and runs as fast as he can, to get into a turn-off, or to his destination, before he meets any one to bother him."

"But have you never any serious accidents from collision? The lads seem to run off in reckless haste."

"We seldom have anything worse than a stunning blow. A lad may be running 'in by' in a heedless, half sleepy manner, with his empty tub, when he is surprised and wakened up by another lad running out with a full one. Of course, the tubs join, and it is the empty one that rebounds; and if the lad going in with the empty tub be not on guard, and have not his arms rigidly set against his tub, then his head is bumped, and almost driven in between the shoulders. You see the barrow-way (a term we apply to the track laid on the bottom slate, before it is cut up for a horse road) is made just high enough for the tub to clear the roof, with a few inches to spare; and the tub being only three feet in height, causes the putter to run in a stooping position, with his body nearly parallel with the roof of the coal seam; for this reason, his head is in a bad place, in the event of collision. But the neck feels the force of the blow, and suffers the most from it."

"Putters are not often injured, I have noticed. How do you account for the fact?"

"I cannot tell. I think, however, that their occupation awakens them to a sense of danger for which they are almost constantly on the watch. They take heed generally as they go in by, that they do not come in contact with those coming out. They have an eye to every place end they pass, out of which a putter with a full tub might pop on them at any minute. Wagons of putters very often get jammed together, but seldom unawares of the putter, if he have a quick ear; and I do not think any one with a slow ear would answer the purpose of a putter at all. He could not escape having his neck broken. But there are no deaf putters, and there are no lame or unsound putters. These could not do the work, which you see is of the most severe kind we have in the coal pit. They must be sound in wind and limb. Woe be to him who is not, and takes up the craft. Very often when boys are scarce, some of those tramps coming along, seeking employment, are tried; but not one in ten ever succeeds or stays. They get on the barrow-way with an empty tub, and they seldom ever reach their destination before they get faint-hearted, and give up putting as a bad job. They meet with the putters coming out, and are glad to get into a roll off, or a siding out of the way of the practised putter, and his jeers."

"The putters do not go to the same places, and get coal of the same miners every day, without changing their routes, or *ranks*, or *sheaths?*"

"Every morning we put in *cavils*, which decide the *sheath* of each putter. Putting in cavils is often done in this simple manner: We take as many *plate nails* as there are putters at a station. We mark numbers on each of the broad heads of the nails, and then put the whole in the cap of one of the lads, and shake them up. Then each of the putters draws a nail from the cap, and the *sheath* of each is determined by the number on the head of his nail. In the table we have shown for example, there are but two *sheaths*, or *sheaths* first and second. First sheath takes the work set in the first column. The second, the work in the remaining column. But with more putters and hewers, this method of placing the work is simply extended. This is the method adopted in the entire Coal Field of Durham and Northumberland."

"There are various ways of drawing the cavils, and when there is supposed to be a preference in the sheaths which can only be slight in any case, there are several ingenious ways of cheating adopted among the lads, in order to get it. For instance, in case of the nails being used, some of the innocents will slyly get the nail, with the number of the rank he covets marked on it, and with his candle heat it slightly on the head, or point, or otherwise, and then he endeavors to be of the first to draw; when, of course, he takes out the heated nail."

"You see by the plan that some of the places are very much nearer to the station than others. This is the cause of another kind of trickery among the putters, few of whom are entirely honest in regard to it. For instance, it so happens that the putters of the first and second sheaths get each a certain number of wagons in some of the near places. There are no means of telling how many each putter actually gets, except the miner takes the trouble of keeping an account in his mind, which he seldom does in a near place, because he is almost certain to get all the wagons he needs, even when work goes slowly with the sets. The consequence is that each putter will get as many as he can in those near places, and watch *eagerly* the filling of the near coals. The miner in the second near place has often his work divided between the two putters of the first and second sheath, and these will seldom if ever pass the place to go farther if the spare tub is filled in this place. The miner is pestered beyond the limits of his patience by the cry of 'Is she full, hinny?' Of course these putters dispute about the number they get in such near places, each one declaring that the amount is less than what he has actually received. It is the reverse of this when a place 'far off' is concerned. This is one of the chief causes of dispute. But if a hewer in a 'far off' place does not get his coal brought out and does not get his share (*shift*) with the other men, the putter must, by an established rule, pay for the deficiency, if it has been on account of his neglect. But the rule is not always strictly enforced, which depends on the generosity of the miner who has suffered a loss by such negligence. A few threats on the part of the miner terminate the business, when all he receives for his pains are the saucy, careless retorts of the regardless boys, who have been the cause of his wrong and justly the subjects of his wrath."

"In case of jumping the track with a full tub, how do these boys proceed to put it on the road, considering them to be so heavy? Ten hundred weight, I see, is the weight they carry."

"Well, you see the wheels are ten inches high, and they are placed so near to each other that a weight of 200 pounds put on either end of a loaded wagon will overbalance it. Now, these boys, after a little practice, can lift this weight where the ground is level, or nearly so. If the tub is on the main headways, he always gets a lift from the putter or putters he blocks in or out. If no putters go the same route, he manages to get his tub on the track, if not by a direct heave, by the aid of a loose plate which he uses as a lever. In such cases, however, he does not spare the deputy; he rates him with all the vehemence of his *pitmatic* eloquence, and hard passionate words escape from his organs of speech. We forgive them, when we remember that we were putters ourselves, and that the severe strain required on the muscular system by lifting till one sees *stars* and scintillations, so excites the mind that a poor fellow hardly knows what he does say, when hindered by jumping a rail. We know that in a few minutes the crying and swearing and passionate exclamations are all forgotten after the *plate* has been replaced and secured. But a great deal of difference exists among the boys regarding their skilful methods of handling a full tub. Some remember every bad joint, or uneven place in the barrow-way after they have passed over it a few times, and running at the top of their speed, a twist to one side or another will enable one putter to pass over a piece of bad road successfully for a whole day, while another putter will not pass over with a single tub without tumbling off the track; and we hear more noise made by the crying and swearing of these than we do from those who use patience and reason to aid them in their physical endeavors. But they all get aggravated at the fact of getting off the track. The deputy, as I have said already, is not spared when a loose plate is the cause. But our best policy is to humor the boys, and pacify them in the best manner we can; it is the only way to get the work along smoothly."

CHAPTER X.

THE HEWER AND HIS WORK.

We now pass into the barrow-way, and go into the pair of drifts called the leading or main headways. These are the places *D* and *E*, Plate II.

The barrow-way is laid on the floor of the seam, or in a narrow cut of a few inches in depth, to allow the tubs to pass, with a few inches to spare, above them. We find the hewer drilling in the face of the main headway's course. His "jud" is curved (undermined), and the side is nicked, to allow the blast to topple over the coal with greater facility. The chief requisites of a good miner are the ability to deal accurately a sharp blow, and to direct each blow of the pick with such judgment as to cause each stroke of the pick to remove as much of the coal as possible. It is evident that, if a miner takes too much of a hold with the point of his pick, the portion will not yield; therefore, a couple of more blows may be required to effect the object. Those blows being the miner's capital stock, if wasted, will be equal to a certain waste of time and labor, and will represent a certain loss which should be avoided. The greatest portion of his skill may be said to consist in the handling of his picks. By constant practice, he soon may become an expert in the use of them. Then, in such coal a little judgment is required in the management of the blasts.

Our Plate VI.* shows the miner at work, nicking his jud, preparing it for the

---

* The Hewer or Coal Miner.—In Plate VI. we find the hewer cutting a niche in his already undermined face of his board-room. He terms the operation nicking his jud. A pick is a tool despised by a number of fine-fingered people, but it is as useful as it is primitive. Without we had those to use picks, it would be of no avail to use the pen. We have presumed that the mine in which our miner works is well ventilated. We have not shown the brattice, usually put into a board-room, to throw the air into the face. This would interrupt the view. But usually, a brattice of thin boards or canvas is nailed to a row of props set close to the rail of the tramway. The brattice projects into the headway's course so as to catch a portion of the air coming into that passage, and it is deflected into the board-room in sufficient quantity to carry off the gas generated there as fast as it escapes from the pores of the coal, as it is literally being dug out by the miner's pick. Therefore, our miner is working by the naked

Roof of Coal

Floor of Seam

Coal Miner nicking his Jud Headway in first undermining the face two depth of The Shivel then a mile is cut. Blast inserts up rived bondles the dust everywhere are substituted for powder where gas in dangerous quantities exists.

Poney Putter

blast. The mining or curving is shown to have been effected in the excavated part of the face, near the floor of the seam. Another view, Plate XVII., shows the board room with its timber and its tracks.

The headways go in the direction of the cleavage, and the coal coming from them forms the wall's-end we read of in the lists of the London coal markets. They are driven from eight to twelve feet wide, and, generally, are pushed forward faster than other places.

In our Plate II. we have shown three pairs of main headways to be in progress on each side of the main drifts or mother gates. In actual mining, the headways are driven in pairs apart from each other at various distances, seldom exceeding two

light of a candle, which you see sticking on the wall side to his right. He takes advantage of the shade to keep his cutting straight; that is, he does not work behind the shade. As soon as the cut is worked back as far as the mining is excavated, a hole is drilled in the opposite corner, and a proper supply of powder furnished as a charge to blast down the "jud." Of tools, the miner has a heavy "breaking-in pick" of four or five pounds, a couple of medium-sized picks of about three pounds, and a backsider of about two pounds. Then he has a nicking pick or two, hammer and wedge, set of drills, and a shovel. His mine furniture consists in the stool ("cracket") on which he sits. To be an expert pick man is to be an expert miner. Some blows must be given with much force; then, by the use of his eyes and judgment, the neat little blows are given to cut off the pieces, to square up the facings where heavier blows are used, to break through them; thus an attentive miner uses more of his judgment, and less of his strength than one who is careless and indifferent, and who hurries and wastes his blows to not one-half so much advantage. The good hewer pays attention to every blow he strikes.

A board-room is given to two men, mates, who, as a general rule, share their earnings. One of them goes to work at 2 A. M., when he is wakened up at the first calling course. He will arrive at the face of his board about 3 A. M. At 9 A. M. his mate will relieve him; when, in most cases, he will have his jud taken off, and nearly all filled up; and, in some cases, will have commenced the second jud to forward the work of his mate (marrow). This is the custom in the board-rooms of the whole workings. In the narrow headways three juds, of three feet each, are generally taken off by the two mates. In the broken juds, where no blasting is necessary, or could be allowed, on account of the danger from their proximity to the "goaves," the mates work together. But each jud has four men; and, in special cases, where coal is in danger of being lost, six men are allotted to a jud; that is, three men work together in each shift; as the day is divided into two shifts, the fore shift and back shift men as they are termed. In the broken juds the coal is much easier mined than in the whole, on account of the crush which is acting on the ends of the pillars being worked out. This makes it possible for two men, during a shift of six hours, to send out fifteen tons of coal from such a seam as the Hutton Seam, whose thickness, we have seen, varies from three to four feet. At this rate, the two shifts of four men each, usually working the double juds, send out from a pillar as many as sixty tons of coal per day. To mine and transport this amount of coal, keeps the shovels as well as the picks constantly in use. A full tub is no sooner away from the face than another empty one is set in by the wiry muscular putter.

hundred yards, and they lay out the mine in panels, in which the boards are worked. The rate of dip very often determines the distance that the pairs of headways should be apart from each other; the more rapid the dip, the less the distance between the levels or headings, because of the greater difficulty of *putting* the coals on a high rate of inclination. The miner's work in the boards is very nearly performed in the same manner as it is in the headways; but the width of the boards may be driven four to six feet wider, and on this account the yard price usually paid in walls is omitted. But the coal works easier when operated directly across its face, and in some cases it is only paid for accordingly.

The miner in the Newcastle Coal Field is not required to set timber under the roof of his own place, nor lay the barrow-way as he progresses. The deputy does this; and you find the deputy begin to take in props and rails every morning as soon as the men (who come in at about 3 A. M., very often within ten minutes of each other) have been allowed to go to their places. For the purpose of transporting rails and sleepers and props, the deputy possesses a vehicle termed a "horny tram," which you will see unloaded in the sketch of our deputy overman, Plate IV. While the putters are at work, it is not so easy a job to take in the props, etc., which are stored at the station. The haste of the putter, who is paid by the score, will not always guarantee the safety of the deputy and the "horny tram," should they happen to meet him when at a full speed run out of a headway's course or mother gate board, where the barrow-way is in good order, and the grade easy. During the day the deputy sets the props near the faces of the working places, and advances the tram roads, as those places progress. So you see the deputy is not an idler, but he performs as much manual labor as any man in the mines. He sets the air doors in the barrow-way, and keeps the air up to the working faces by putting up the brattice and temporary air-stoppings. He takes the rails out of the old places after they have "holed" into others, and *draws* out the props of the boards after the rails have been taken out. The drawing out of the timber is done as over-work, and paid for by the score, or at the rate of twelve cents for twenty props. But the hours of the deputy are not long, if they are busy ones. At ten o'clock A. M. he is relieved by the back shift deputy, who, from that moment, takes the cares and responsibilities of the district upon his shoulders, and he remains at at his post till all hands have left at quitting time, about five o'clock. He then sees that all air doors are shut, and that the air currents are working satisfactorily. The miners also work in two shifts, and the fore-shift miners are relieved at the same time that the fore-shift deputy is; but the putters and drivers and onsetters and road men, together with the men on the surface, work twelve hours per day. Of course, the labor of these men is not so excessive as is that of the miner, excepting in some cases the labor of the putter, which to perform rapidly requires great physical force and endurance. We have followed on through the daily routine practised in a district working the "whole" coal at the dip of a mining property, the boards being driven to the dip, whence the light gases drain off naturally. We shall now go from the shaft up a self-acting incline, and examine the work as it progresses in the *broken coal*. But there are some attendants whom we have passed unnoticed on our way, who deserve a passing word at least. The boys who couple the tubs at the top and bottom of the engine plane and give the signals to start the engine sets have responsibilities quite important for their ages; boys about fourteen to sixteen performing these services. Then there is the door-keeper as shown in Plate VII., and that little horse-driver of twelve or fourteen, with his horse of about sixteen hands, who is not to be jeered at, as his picture in the plate will tell you. He works off the engine, sets and takes ten to twenty wagons at a time, and acknowledges as superiors only the "wagon-way-man" and the *overman*, an officer we shall accompany just now through the broken workings. But the wagon-way-man is responsible for the good keeping of the horse roads, and for the road in the engine plane, together with the rollers on which the wire-rope, working the engine sets up and down the plane, runs. The wagon-way-man is usually one of the *handiest* men in the mine. He must be an expert worker and have good judgment, and pay strict attention to his roads.

CHAPTER XI.

THE OVERMAN—SELF-ACTING INCLINED PLANE.

THE overman is an important personage, filling a very responsible position. This officer we find, after he has been in his cabin and stripped off his superfluous clothing. He is in his drawers, and a sleeved vest of blue flannel. You see the breast of his white flannel shirt, his clean blue stockings, and strong shoes. His leather cap and "yard wand" belong to his mining outfit. (See Plate IX.)*

---

* THE OVERMAN (Plate IX.).—Among the executive officers of the coal mine are two, mention of whom has not been made in the text, and we must not omit them here. We refer to the Master Wasteman, and the Master Shifter. The one has charge of a few men who keep the air courses open, and who make stated excursions through them. It is remarkable that all master wastemen are more or less stout. One would suppose that he was purposely selected for the special duty of getting through the airways; for a corporation like his going through an air-course, is a guarantee that a large area of said passage is free, at least as much of it as will correspond to our wasteman's ample proportions. The overman has generally passed all the grades, except in some cases that of master wasteman, of whom we cannot say anything here except that he is stout, and that the duty assigned to him is the care of the *return air-courses*. Consequently, he is always out of sight until time to quit, and time for dinner takes him home, as he is always a man of the "second," or "boys' calling course." A master shifter the overman has often been; but it is not absolutely necessary that he should have been such to fit him for the post of overman. Another officer, too, who belongs to the underground department of a colliery, is the master sinker, who very often follows this profession as a steady occupation through life. Our overman may have been a master sinker, who is generally considered a very clever fellow, and who, after the sinking of a new shaft or "winning," is often retained in charge of the coal mining, especially in new coal fields. But very generally our overman has come up from the trapper; and the office preceding that of overman is very generally that of back-overman, whose duties are so similar to those of a superior deputy overman, and who approaches so nearly the chief overman, as not to require a special sketch or description.

The duty of the overman is to attend to the workings, and arrange the men at their work. He consults, at least once a day, with the underviewer, to whom all contemplated changes are reported and explained. He receives the reports of the deputies of each district every day, and never fails to see and interview them, when all important points are discussed. In this way, all parts of the mine are brought constantly before our overman. In some special cases he has studied the higher branches of science, and become an expert scholar and VIEWER or MINING ENGINEER, and here we shall leave him at the *top of an important profession*, and we shall enter his parlor or comfortable sitting-room, and hear what he has to say to one of our American fraternity on a *tour professional*. You are ushered into the presence of an intelligent gentleman, of distinguished ability. He suggests a glass of wine which he

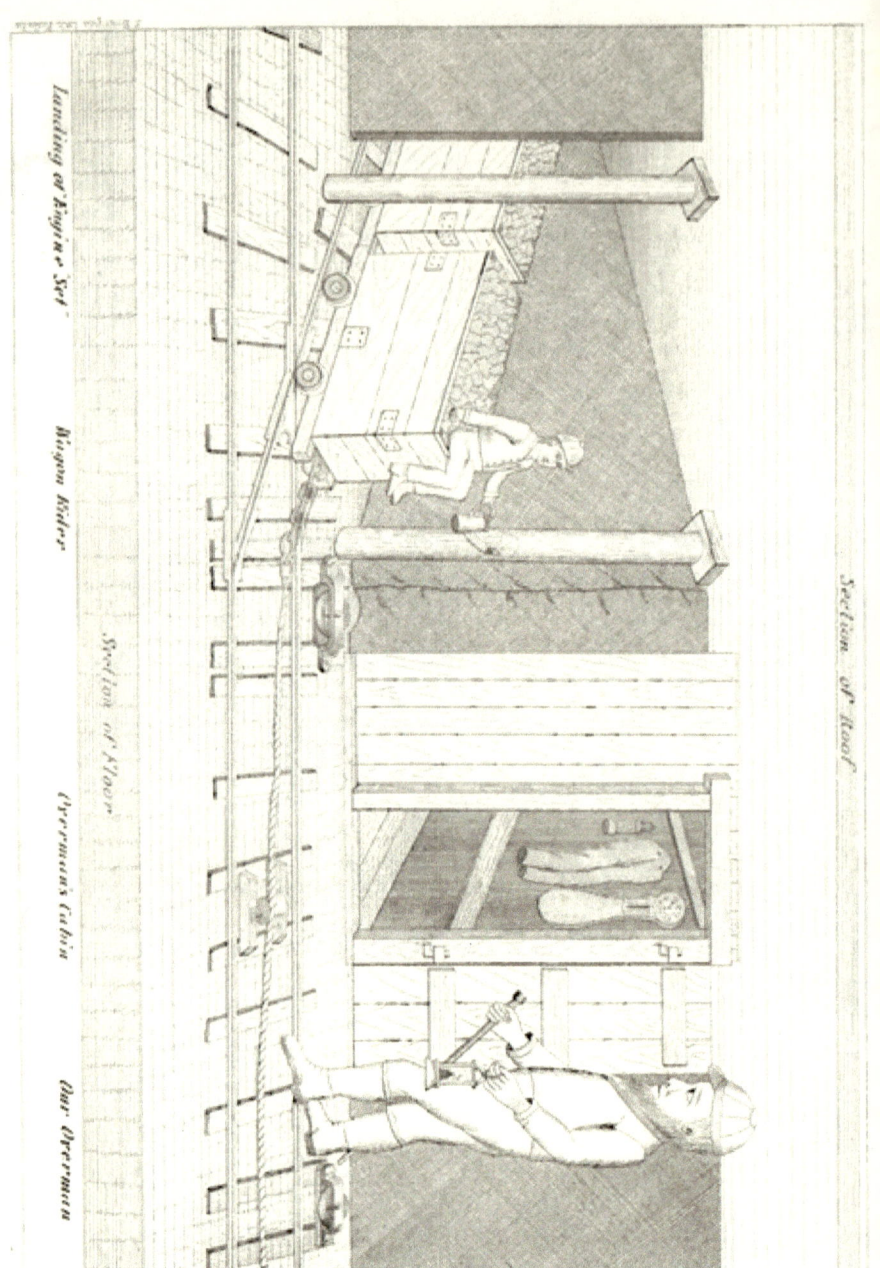

The man before us is fifty at least. He has "kept" a door and a switch, and been a "way cleaner" before he was twelve years of age. Then he was a driver, a helper—declares is an old custom, becoming obsolete. You decline the civility with respect and many thanks. But you take a cigar, when it is offered. Your friend takes his long clay pipe and stuffs it with the best of Virginia's produce, and he enjoys his pipe, while the wine stands untouched between you.

"You have come from Germany," says he, "and you must have visited the mines of Westphalia, Belgium, and France. It is the best school for study," he continues, without allowing you time to answer the question he has asked. "We English are in the main too conceited to travel with a purpose. But we are changing for the better, in regard to this matter. We are beginning to seek practical knowledge in the Continental districts. We have become so accustomed to supply the world with mining engineers that we may have been jealous of our neighbors. Into France, Germany, and everywhere else, we have sent trained miningmen, except, perhaps, to America; we may say we have sent our pupils to all parts of the world. Some have gone to America on speculation, but have returned dissatisfied. They find the 'miner boss' and ordinary surveyor installed in the places they are trained to fill. We have no right to find fault with an arrangement which for the present time seems to give satisfaction. At present, judging from accounts, it is hard to say whether the miner boss or superintendent, or the surveyor or the master machinist, or outside boss will become the mining engineer of America. This is your own affair. Isaac Lowthian Bell, one of the members of our Institute, has iron interests in the southwest of your country. He has written his report, and spoken much of his visits to your coal regions. You have fine natural resources; mines almost vomiting their riches into your very laps. Your gold and silver jingle out of the mountain gorges. Your copper and lead have overflown from the crevices of the rocks. Your oil wells spout into the air, and illuminating gas ready-made, or nearly so, flows unheeded right in the midst of some of your large towns. Why the best coal in the world has tumbled into your wagons, just at the mere touch of the 'starter's' or miner's crow-bar. You have so far scarcely required the aid of the mining engineer, and from what we learn, you get along, as best you can, without the aid of a mechanical one, which in a new district is a matter of still greater difficulty. The managers you have seem to be very jealous of mining experts and professionals. They seem to hurry and push ahead a lot of work hardly worth the doing. There is this difference, you will always observe, between men who understand their business, and men who do not. These latter push and fret, and take a great deal of trouble to do improperly that which would be of no trouble to those others who have taken the trouble to get started at the bottom of their trade, and get up to the top of the ladder. They go about their business as if they had nothing in hand, and as if important jobs gave them not the least trouble."

The fire has gone out of your own cigar while your host has been speaking, and you relight it; then you strike a match, and hand it to your host who, thanking you, lights his pipe again. You smoke for a few minutes, without speaking. Then you talk, as an American student well knows how to do, in a creditable manner. You amuse your host with a description of the Continental mines and miners; and in some things you literally instruct him, particularly on the application of electrical apparatus employed in the French and Belgian mines for signaling and for lighting the main portions of them.

You have smoked out your Havana through your descriptions, and venture a remark on English practices and prejudices in mining. But he continues his remarks aptly: "Our systems of mining and transportation here in the north have always been in advance of the other districts, as a whole. Of up, a half-tram, a timber-leader, or "deputy's clerk," and a putter. At twenty he was a hewer; at twenty-four a deputy; at thirty a back overman; at thirty-five a *maister shifter*, and at forty an overman. He is satisfied with his position, and being a master of it, he feels no anxiety concerning his heavy responsibilities. He looks like one ready for any emergency; and nothing could take place inside of a coal mine that would take him by surprise. We have nothing to do with his physique, which is such as no man need be ashamed of. His dress belongs rather to his profession than to his person. The flannels are a protection to the miner, in case of slight explosions of gas. The parts of the body covered over with flannel are not often burnt, the flannels being bad conductors of heat. We walk on to the west of the shaft, and go along the shaft siding. A set of full tubs have just been delivered from the self-acting inclined plane—a road worked on the principle of gravitation. The shaft boys are running up the empty tubs from the bottom of the shaft to form the set of the "*incline*, and as each tub is *joined up* it is coupled. Thirty tubs complete the set, and here is the last one just coupled up. The lad pulls down a lever fixed on the side of a piece of timber, and lets it spring back

late, we have substituted the long wall of Derbyshire for our board and pillar system of getting the coal, which, in many cases, is to be preferred. This is the principal change which has taken place. We have substituted iron for wood at some of our hempsteads. You can see from the window behind you our elegant structure, built of trussed iron. Mounted over all are these fine wrought iron pulleys which are twenty-three feet in diameter. They stand seventy feet in the air on their light, elegant, yet strong system of iron network. It was mainly suggested by an English mining engineer who had served an apprenticeship to mechanical engineering in your country, in the works of Cooper & Hewitt, if I remember right, of Trenton, New Jersey. He came to me from Mr. Crawshay, of the Gateshead Iron Works, about the same time that Mr. Field, of New York, came to us to consult on the laying of the second Atlantic telegraphic cable which we took in hand. Peter Cooper, of the firm I have mentioned, was the president of the cable company. He knew our iron masters here, the Crawshays; so your engineer and I became good friends. He came with an extraordinary offer. He proposed to erect his iron hempstead, with the pulleys mounted on it, at the cost of the Gateshead Iron Works, and asked that, if it should suit us, we should pay the cost; if it did not please us, he proposed to tear it down, and remove it. We did not accept the proposition at once, but took the drawings, consulted Mr. Crawshay, and after a year or two, put up the improvements as you see them. We objected to the plans then more on account of having our business interrupted, than for any other reason. We added to your American idea of truss-work our English ideas of a screen made to screen out the small coal, and to weigh the large coal; and taking advantage of this plan to find out the miners who sent out the largest proportion of lump coal, to whom we paid premiums. We found that such a system benefited the proprietors, as well as the best of our miners, if not the whole of them."

again to its original position, and the signal **to start is thus given.** The bell wire shakes and rattles, and shortly after the rope on the **front of the set pulls up in line** with the first tub on to which it is hooked, and the rope pulls, **and the *set* starts** steadily off up the plane after the wire rope; and after the set has disappeared, **we** hear it rumble up the plane till the receding sound is lost in the distance. Waiting here **for a short time, we hear the** sound of wagons again on the plane **approaching.** Then the set rolls into the siding, and comes to rest; when we leave the shaft boys busy uncoupling the full set, and getting another empty set ready, and we go on up the incline.

The *inclined bank* going west, and to the rise, is very similar to the *engine back* we have seen going on the dip to the east. There is a single track, in the middle of which there are rollers to carry the wire rope. This single track goes on to " meetings," a siding at which the full set coming down passes the empty set going up. The tongues are fixed and open, and the empty wagons go into that track, to which a pair of peculiar switches directs them. The empty set goes on into the siding, and after it is well clear of the full set which it has met, runs on a track formed of three rails. These three rails form two tracks, but the middle rail is common to both of them. The full sets going from the siding at " meetings" into the single track below " meetings" shift the switches automatically, so that the empty set, coming up the plane, goes into that track of the siding which the full set has just left. Where the curves of the road have a tendency to draw the rope out of line, there are sheaves which retain it in the middle of the track. We should have remarked the same thing in the engine plane. On each side of the inclined plane drift, we find the cross-headings well stopped up by the substantial brick stopping, well cemented and plastered, to prevent air leaks. This denotes that there is on each side of this main drift to the western boundary of the mine, a return aircourse collecting the returning air-currents, coming from the north and south sides of the colliery's workings.

We find on the head of the " inclined bank" that the three rails run into a single track, and the single track branches out at once to form a siding of two tracks, one for the full, and another for the empty wagons. Here we also see that the sets coming up the inclined plane are divided into horse sets, and distributed to the various districts whose products are brought to the head of the plane.

The machinery that "brakes" the sets up and down the plane is simple in plan and detail. A system of sheaves (three in number), six feet in diameter, around which the wire rope of the plane partly passes, constitutes the winding or lowering machinery. The spindles of the sheaves are set vertically, and the grooves of the three large sheaves are in the same plane; and they are provided with a brake operated by a hand-wheel, pinion, and racked lever.

A strong boy operates the brake, and this boy is termed the brakesman, who regulates the speed of the sets. Another boy who couples the wagons and drives a trained horse to start off the sets, constitutes the whole force required to operate the sets at the head of the "inclined bank."

PLATE VII

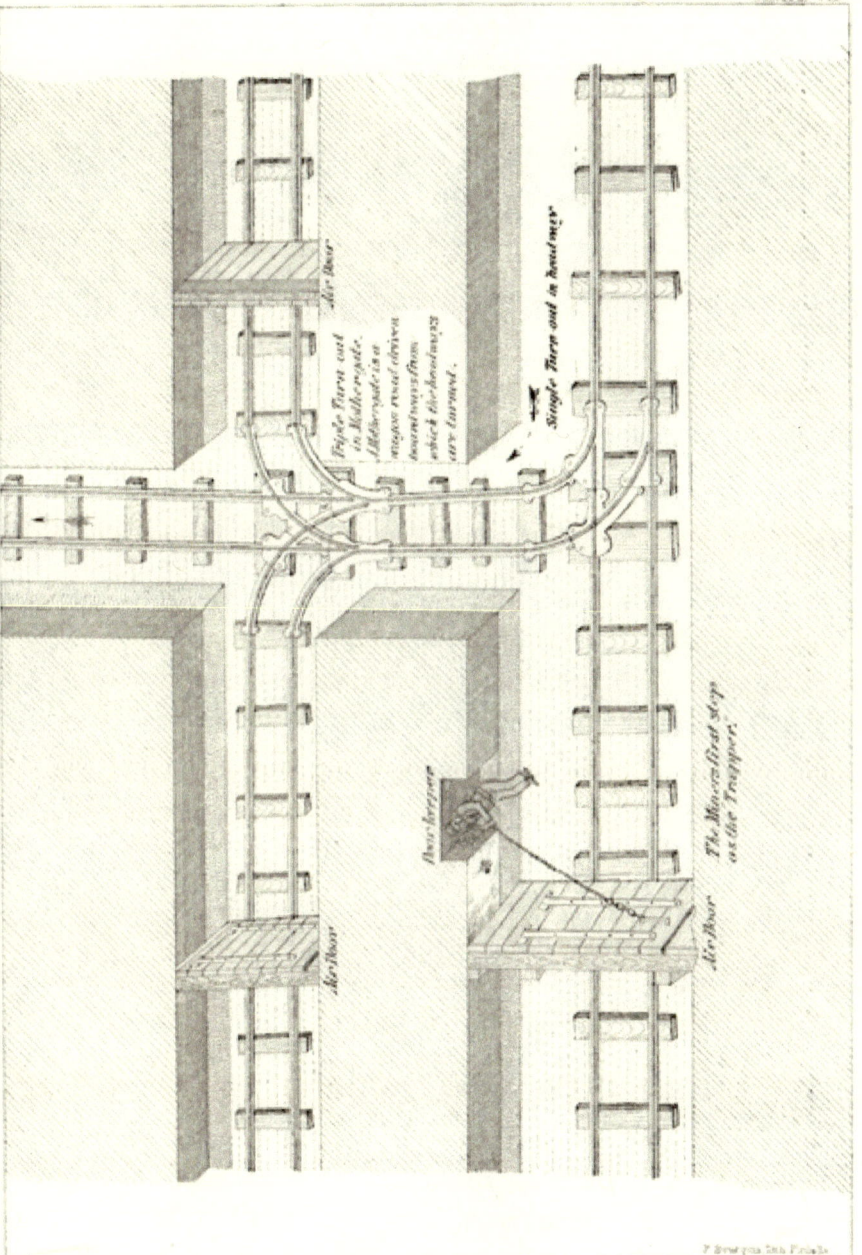

PLATE X.

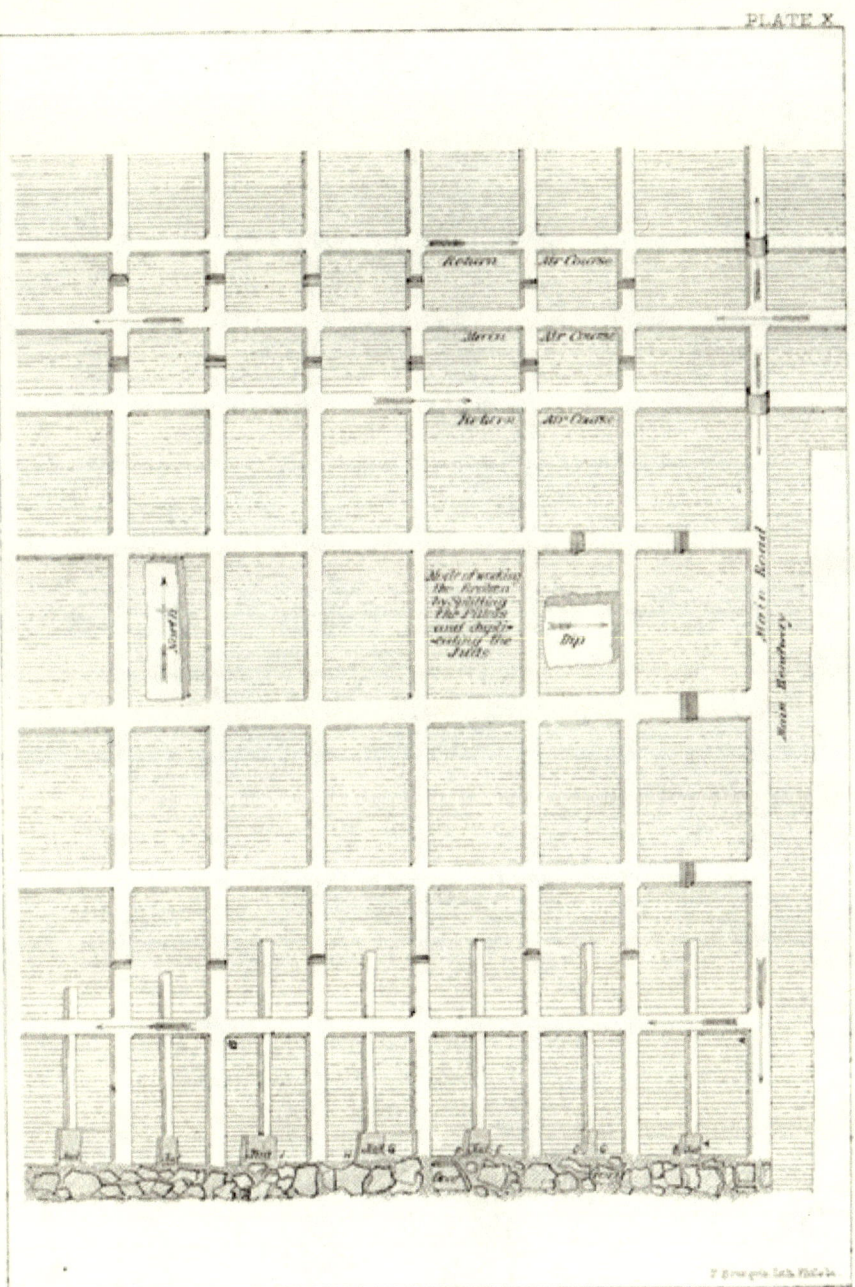

CHAPTER XII.

"BROKEN COAL" WORKED.

We pass into the horse road, leading to the district south, where we find the *broken coal* being worked out. Here the overman will explain to us the various methods adopted in taking out pillars of coal, and his instructions need not be marked by the signs of quotation. The horse road is very often the main headway of a district. In Plate X. we have marked it as the main road. It has been of late years an object in the ventilation of a mine, to avoid, as much as possible, air doors. You will see that in our main tracks, as far as we have gone, we have no air doors; and when there is an abundance of fresh air carried forward, there is no necessity for them (for air door, see Plate VII.).* It is only the mines inadequately

---

* Plate VII.—The miner's first step represents the little "Trapper" at his door in the horse road. Trapper is a term derived from trap-door, which was formerly used to designate an air door in a coal mine. The scene shows the little fellow in his niche or "hole" cut in the rib of the wagon road headway. The door at which he is posted in this instance, is a wagon-way door. When such a door is situated near the switch of a forked track, the duty of switch-keeper is added to that of door-keeper. In our sketch, we have allowed him a candle, which he sticks to the coal-rib in a piece of tough clay. As a general rule, the trapper works in the dark, and acts in obedience to the communications he receives through his auditory senses. When he hears a wagon rumbling towards his door, or the step of a foot passenger approaching, his duty is to pull on his door string, one end of which is secured close to him, and open his door. As soon as the wagons or passengers are safely through, he allows the door to fall shut.

Air doors are made to fall shut by gravity, by a certain mode of hanging them. The frames are not exactly perpendicular; the king post leans at the top a little towards the track on the one hand, and as much away from the door on the other. Some of the main doors, which happen to be located in passages connecting the intake and return air courses near to the shafts, are subjected to a pressure amounting at times to as much as one hundred and fifty pounds, which requires a little tact to handle them and jerk them open. These doors are generally relieved of pressure by placing two doors (double doors) in the passage, within a short distance of each other, yet far enough off to allow a horse and his set to stand between them, so that the two doors have no need of being open at the same time. This is a plan also used to prevent an undue escape of air through such passages.

By a skilful application of the Regulator and Air Crossing, the use of air doors has been abandoned in some of the best ventilated coal mines. We have here reference to *main doors* in *main passages*. In passages used temporarily, near the face of the working places, doors may be used without

supplied with air through the insufficiency of air ways that are termed "FIERY." Where there is not an abundance of air, the lame method of using doors to force the little air in circulation here and there and everywhere, is resorted to. Plenty of capacious passages will furnish an amount of air, in proportion to their capacity. If the air is forced with a high velocity, it will rush through the nearest outlet, and a small opening or leak will waste a large amount of air.

The overman tells us this as we pass through the main headway. A horse and train (set) of loaded wagons pass out as we go in, and the driver, a sharp lad of thirteen, sits rather comfortably on the *limmers*, holding a safety-lamp to light the horse on his way, as we have seen him in Plate VIII.

We come next to the station which, as in the "whole coal," is only a siding in which the horse sets are collected. This is a large district, extending from A to B, the coal above belonging to a system of workings on a higher level, and in it are ten "juds" from each of which thirty tons of coal may be sent daily. In each jud two men work at a time, and during the day of twelve hours these men change, or are relieved by a second shift. On account of the downward action of the roof, the coal is easily mined with the pick, and during his short shift of six hours, the miner never ceases his work. The pick to loosen the coal is succeeded by the shovel to fill it into the tubs, which are quickly changed by the half naked putter, with the drops of perspiration falling from his chin.

inconvenience, and with tolerable safety. In horse roads all doors should be dispensed with, if such a feat is possible.

In place shown in this sketch it is desirable to turn the air up the *board* driven off the headway. To do this, in a continuous stream, it is necessary to put doors in the lateral passages also, as shown by the doors placed in the back headways. In actual practice, doors are not often placed so near to each other as they are shown to be here; but they are quite frequently placed relatively so. I should advise those used to ventilate or force air by the use of air doors, to study the merits of the Regulator and Air Crossing, and solve the problems in ventilation to which they may be applied.

We have taken advantage of this sketch to illustrate the method of tramway so extensively used in the mines of Durham and Northumberland. It will be observed that the single turn-out is made up of three pieces of casting, besides two sweeps and one straight piece. They are for tracks which branch off at right angles from a main track. They are used more in the barrow-way than in the horse roads, in which latter the ordinary switches and frog crossing are used to lay all turnouts and sidings. The triple turnout is very rarely used. Complete, it consists of five pieces of casting, besides the two pair of sweeps. By dispensing with the two upper pieces of casting, representing each a crossing, we have remaining the ordinary *double turn*, so useful in working-off the juds in the broken.

The Waggon and fastners as used in the mines of Northumberland and Durham England

The Horse's second step
The lower

Front of Waggon wheels Lower
Axle box and Strap
Side of Wheel

"BROKEN COAL" WORKED.

There are twelve putters in this district and the deputy's "plate layer" or "timber leader," an attendant who is indispensable in the broken workings, has his troubles among them, fixing the *tracks* when there are so many together. We shall place the work as it is usually done, in a broken district, twice a day by the deputies. The letters on the blackboard refer to the letters of the Plate X. denoting the several juds. We have excluded those places driven outwards to split the

Juds.	Number of tubs to each jud.	Sheaths or ranks of Putters.										
		1	2	3	4	5	6	7	8	9	10	11	12
A	42	17	18	7	..	..	..	..	..	..	..	..	..
B	42	..	..	10	18	14	..	..	..	..	..	..	..
C	42	..	..	..	..	3	18	17	4	..	..	..	..
D	42	..	..	..	..	..	..	..	14	17	11	..	..
E	42	..	..	..	..	..	..	..	..	..	7	17	18
F	42	..	..	..	..	..	..	..	..	..	7	18	17
G	42	..	..	..	..	..	..	..	14	18	10	..	..
H	42	..	..	..	..	4	17	18	3	..	..	..	..
I	42	..	..	11	17	14	..	..	..	..	..	..	..
J	42	18	17	7	..	..	..	..	..	..	..	..	..

12)420
2)35
17½

pillars; but usually the coal to come from them is *placed* along with the coal of the juds, and they are arranged in their order, according to their relative distances; that is, those nearest to the district siding coming first in order in the column in which the miner's work is placed. The number of wagons allowed in such places would not be more than three-fourths of the number allowed for men working in the juds, and the price of the mining of such coal would be increased at a rate to make the wages of the pillar splitters equal to the wages of the men working off the juds.

By reference to the above table, which corresponds to the blackboard nailed up in the district, we find that there are ten juds and twelve putters. The work placed is for one-half a putter's day or one whole shift of miners. Two men in each jud are allowed forty-two wagons, which are termed two score, the extra tubs being allowed to make good all bad filling and waste product. The total summed up and divided among twelve putters gives each boy thirty-five. Subdivided by two gives the putters, either seventeen or eighteen "near" coals, with the same number of the "far-off" coals. By inspection of the blackboard, it will be seen how the odd numbers are divided. The first putter gets 17 tubs at the jud *A*, and 18 at the jud *J*. The rest will be seen as they succeed each other. It will be seen that several of the putters go in four different places; and there are two of the juds *C* and *H*, in which four different putters go; but the arrangement is so well understood that there is seldom any confusion in either the operations of *putter* or *hewer*.

We will, with the reader's permission, ask the overman a few questions concerning the working off of the juds.

"What are the main dimensions of the pillars of this district?"

"The coal being highly charged with gas, we have done well to have them made so large as they are: namely, sixty by ninety feet. These you see are split headways, that is, on the end of the coal. The juds in each pillar are duplicated; one is worked off to the dip, the other to the rise, and the work is pushed to the utmost to preserve the coal from the crush following up from the 'goaf.'"

"How long does it require to work off a jud?"

"Let me see:" and the overman makes a mental calculation. "In this district the juds only last two days; and a jud of five yards in width will yield above one hundred tons of coal. As we take off about fifteen yards of each pillar weekly, this range of pillars yields in the neighborhood of eighteen to twenty-four hundred tons per week in the aggregate."

"Then the yield of coal from a district working off the broken coal is quite an item in the produce of your colliery?"

"With the places splitting the pillars taken into the calculation, we do much here to keep the shaft pulleys running; we get three hundred tons of coal weekly from those in addition to the amount we get from the juds."

"Have you much trouble with the crush of the roof as it acts with its immense leverage over the goaf, pressing on the ends of your pillars as if on a fulcrum?"

"We have the usual crush; but we keep the pillars and juds in line with each other, or as nearly so as we can, to offer as strong a line of resistance as possible. When any small number of the pillars are left a little behind the others they

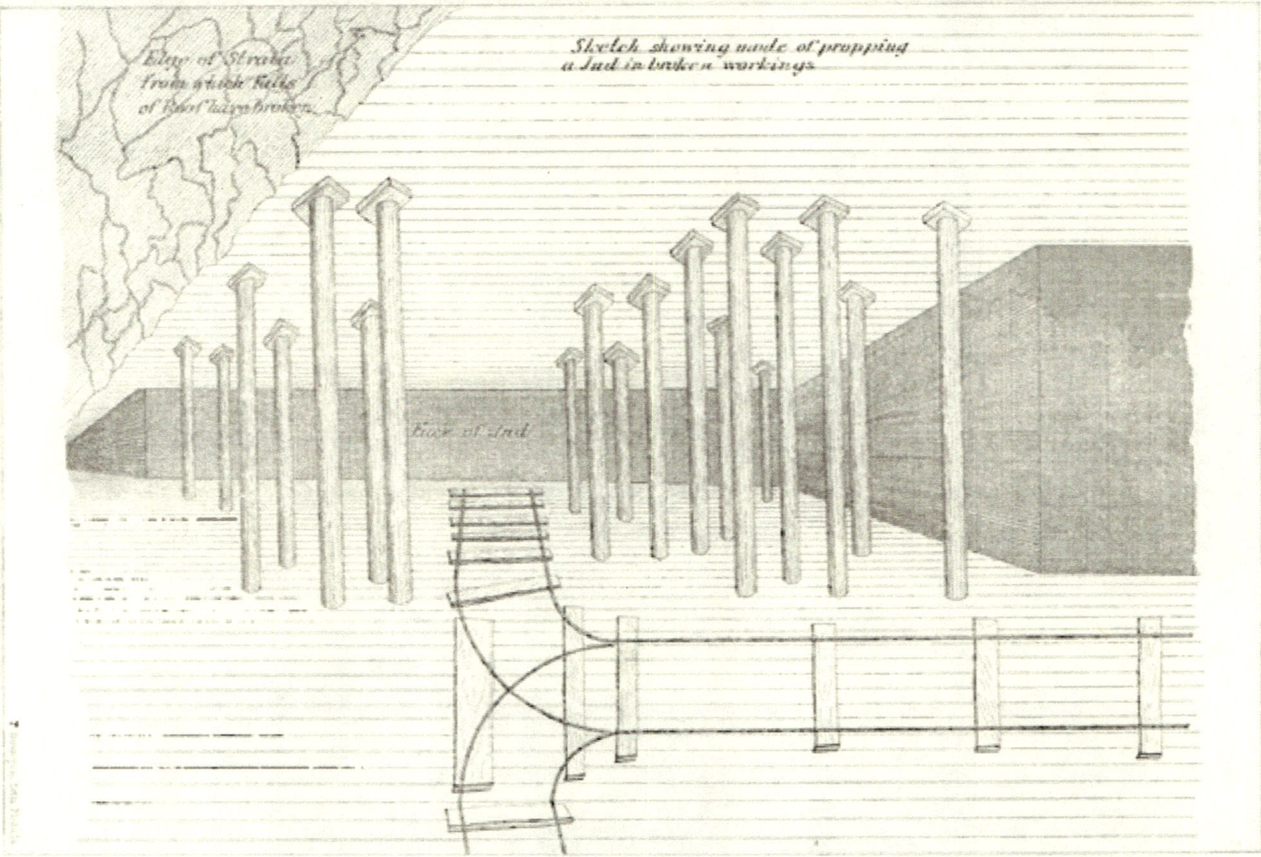

PLATE XI.

Sketch showing mode of propping a Jud in broken workings.

PLATE XII

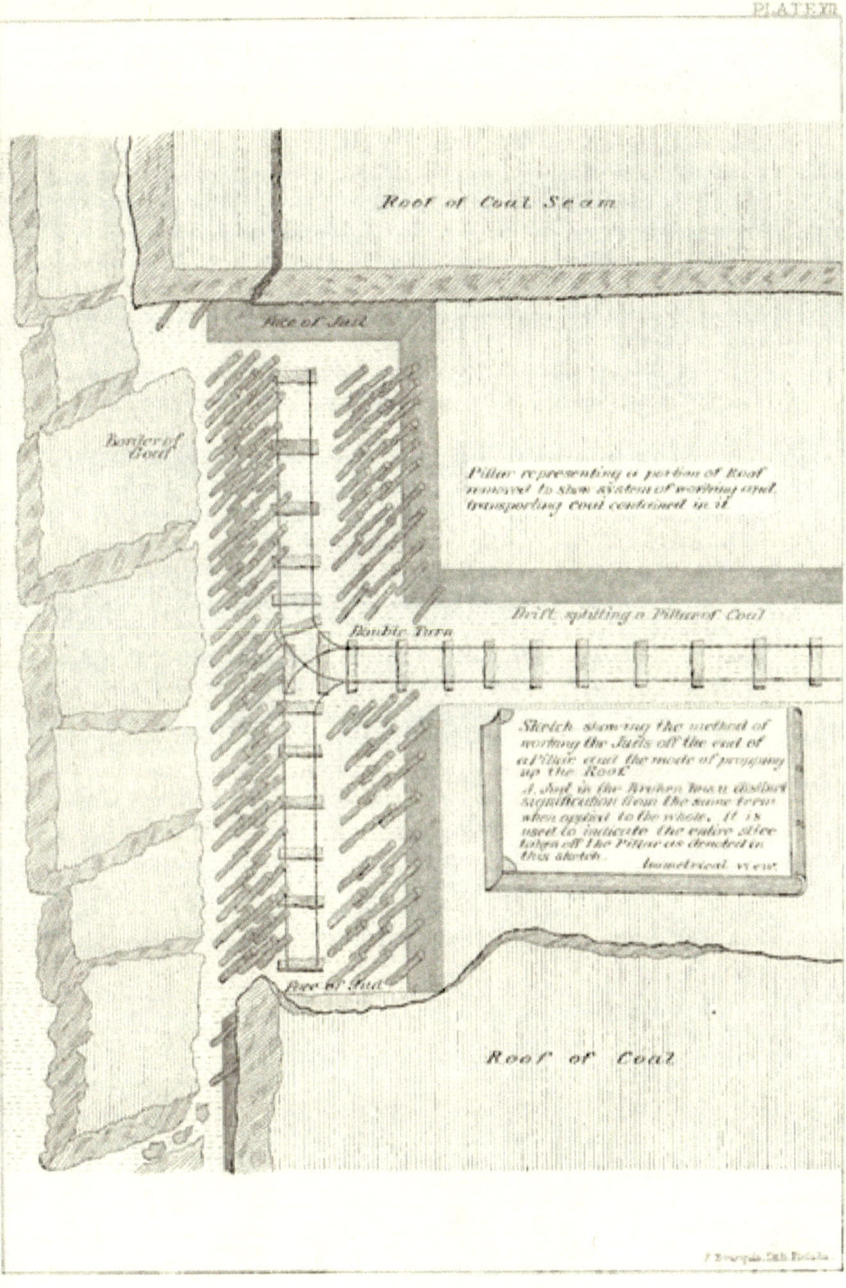

Roof of Coal Seam

Face of Jud

Barrier of Goaf

Pillar representing a portion of Roof removed to shew system of working and transporting coal contained in it.

Double Turn

Drift splitting a Pillar of Coal

Sketch shewing the method of working the Juds off the end of a Pillar and the mode of propping up the Roof.

A Jud in the broken has a distinct signification from the same term when applied to the whole. It is used to indicate the entire slice taken off the Pillar as denoted in this sketch. Isometrical view.

Face of Jud

Roof of Coal

F. Bourquin Lith. Phila.

suffer much from the crush, and give us much trouble in keeping the roads open to convey the coal away from them. As it is, we experience little trouble. By thus keeping the pillars in line, and working off the juds rapidly, the roof gets down with such facility as to avoid a great deal of the creep caused by a roof that is held up by a stump of coal left here and there at too many points in the goaf. Stumps of coal at some places (I mean at other collieries) are left by carelessness, and they have the effect of deflecting the pressure of the roof over the adjacent pillars to a great distance. This causes the juds to be worked off *totally* with much difficulty. The great object to be aimed at in working the broken is to get the coal all out before a jud comes on, that is the weight of the roof." (See Plates XI. and XII.)

"You mean the weight of the roof in the jud coming on to the props. But when this is made manifest by the breaking of the props, you must draw out the timber?"

"Yes; but only at the last moment, if the work has not ceased for the day. We have sometimes to draw out the timber before quitting time; but we always aim to get the jud off about time to quit, so that the back shift deputy can draw out the props at night. Then the *shifters* generally get the new juds ready to start off next morning."

"The drawing out of the props is attended with much danger?"

"It would certainly be as dangerous a job to take out the props of a jud as it apparently is, did not the deputies set up every prop themselves. You see the props are set as closely together as it is deemed necessary, and the deputy while drawing the props does it expeditiously, keeping himself secure among his forest of props in the mean time."

"Begin, please, and describe more minutely the drawing of a jud, and we shall follow the interesting operation through, till the props are piled in the end of the nearest board ready for further service."

"The time is after the last tubs of coal have gone out for the day and the miners have put on their clothes and gone out; then the deputy—with his timber lad, and, in some cases, a hewer he keeps to assist him—comes in with his axe—a tool made with a hammer face on one of its extremities. He commences operations at once, beginning at the innermost row of props. As he knocks out one after another the boy picks them up and places them on his tram; and when the tram is loaded, it is run out to the nearest board end, into which the props are carefully stowed with their ends pointing outward so that they may be counted, as the drawing is paid for at the rate of sixpence a score. By the time the tram returns a number of props have been thrown out into the tramway, and the lad begins to load his tram as quickly as possible. As the tram-plates are pulled out at the same time, these with the sleepers are loaded on the tram, and they are also stowed away to be ready for immediate use. But as there is in almost every case of the drawing out of a jud on the rise another to be drawn out on the dip, it often requires two deputies, the one to aid the other, or each one to take a jud and draw out the props simultaneously in each jud, each deputy being provided with an assistant of some kind, either a man or strong boy."

"When the props are all out will the roof get down at once?"

"Sometimes the fall takes place before all the props are out, and the fragments of roof break to pieces at the deputy's feet. In such a case a few props are unavoidably lost. Then a good man will stand at his post till the roof breaks about his ears and feel quite safe. But when the final crash comes you will always see the deputy in a safe place. He generally has a correct idea of the line at which the roof will break and fall, and he makes it a point to be beyond this line when the roof thunders into the vacant goaf."

"Quite a number of props are set in the juds before they are worked off?"

"Out of a double jud, the deputies draw in about thirty minutes very often fifteen score."

"You have more than one set of deputies in such a large district as this?"

"There is work enough to keep two deputies busy in each 'shift.' At this rate we employ a deputy to each one hundred tons of coal mined daily. But, as you see, we do not allow the hewers to handle any material; they neither set props nor lay rails as they do in other parts of the world. We have the division of labor system more thoroughly carried out than our brothers in other countries, and on this account we mine more coal in less time with an equal number of men. The deputy I consider the most serviceable man in our coal mines. Everybody calls for him when he is in trouble, except, indeed, the driver in the horse road, who calls for that worthy the '*wagon-way-man*.' The hewer, meeting with a piece of bad roof, sends for the deputy by the first putter who comes into his place afterwards. If the face is too far away from the end of the barrow-way, the hewer says to the putter running out with the loaded tub: 'Lad, send the deputy in to lay a length of plates,' and then he goes on with his work of *skelping the coal*;* and after the deputy arrives to lay the length of plates, he can hardly wait to give the deputy time to fasten them to the sleepers, which he does at about the same rate that a man putting shingles on to a roof, fastens them to the shingling lathe. But it is the putter who calls the loudest for the deputy when a 'turn' gets loose or a plate gets astray in a piece of main barrow-way. The plate layer will not give the putter satisfaction at all times. You see that the rail or barrow-way plate is a short one of four feet, while the wagon-way rail differs from it in length only, being a long rail of sixteen feet and weighing twelve pounds to the yard. Each barrow-way plate, therefore, weighs sixteen pounds, and being supplied with two holes at each end, it is quickly laid and secured to the sleepers."

"You work entirely with the safety lamp in the broken coal?"

"As a rule we do. We never see any gas here, because the air in circulation in this district amounts to thirty thousand cubic feet per minute. The yield of gas, as a usual thing, is not more with a low and rapidly falling barometer than three hundred feet per minute. But it is not less under the best conditions of a rising and stationary barometer than fifty feet per minute. You see the variation is great, and we have enough air coming into the district to keep it safe from an explosion if the yield of gas were four times as great as it is. But we contrive to keep the air safe from admixtures of dust, and for this purpose we send water into the horse roads to moisten the dust in the wagon-way. This is done by an automatic sprinkler, which operates only while the 'water wagon' is in motion."

"There are a number of acres of coal pillars worked off in your mine. If into this large space gas were to be forced through a rent or an opening, and make its way into the goaf this space would form, what would be the consequences?"

* Skelping is a term hardly used in the north when the coal is cut with the pick, instead of being blasted with an explosive.

"Simply the gas would force its way into the air skirting the goaf, and very likely an explosive mixture would be the result at that point only. But I have an idea that where the coal is well cleaned out, and the roof let down evenly, no very serious escape of gas is possible; because in case of an upper coal seam the roof will not break to pieces in falling into a seam of coal of this thickness, or one thicker than five or six feet. Beyond a certain height, very limited, the strata will sink *en masse*, and no break will occur. Below, on the bottom, the falls of roof allow the superincumbent strata to sink, and the pressure soon becomes the same on the underlying strata that it was before the coal seam was worked. Except we strike a fault running through a strata, I am not concerned about the gas coming from other seams of coal than the one in which I am working. A fault, running through the measures, if it is not charged with water is very likely to act as a drain to let out the gases of the coal measures, in some cases extending to the surface of the earth. And some of those faults allow much of the gas to go to the surface if water does not clog up the channels and prevent its escape, which, however, is more of a rule than an exception. But, as far as the yield of gas from the pillars is concerned, we know almost how much to expect and when to expect it; and if we are not prepared for our old enemy of the mines, in ordinary workings, it is our own fault; because we have had a great deal of varied experience with it, and have applied a number of tests to it; according to the universal conclusion of our most learned in such matters, the only way to conquer it is to watch it and provide it with an easy, *airy* carriage to hand it politely out of the mines and into the atmosphere at the top of the upcast shaft."

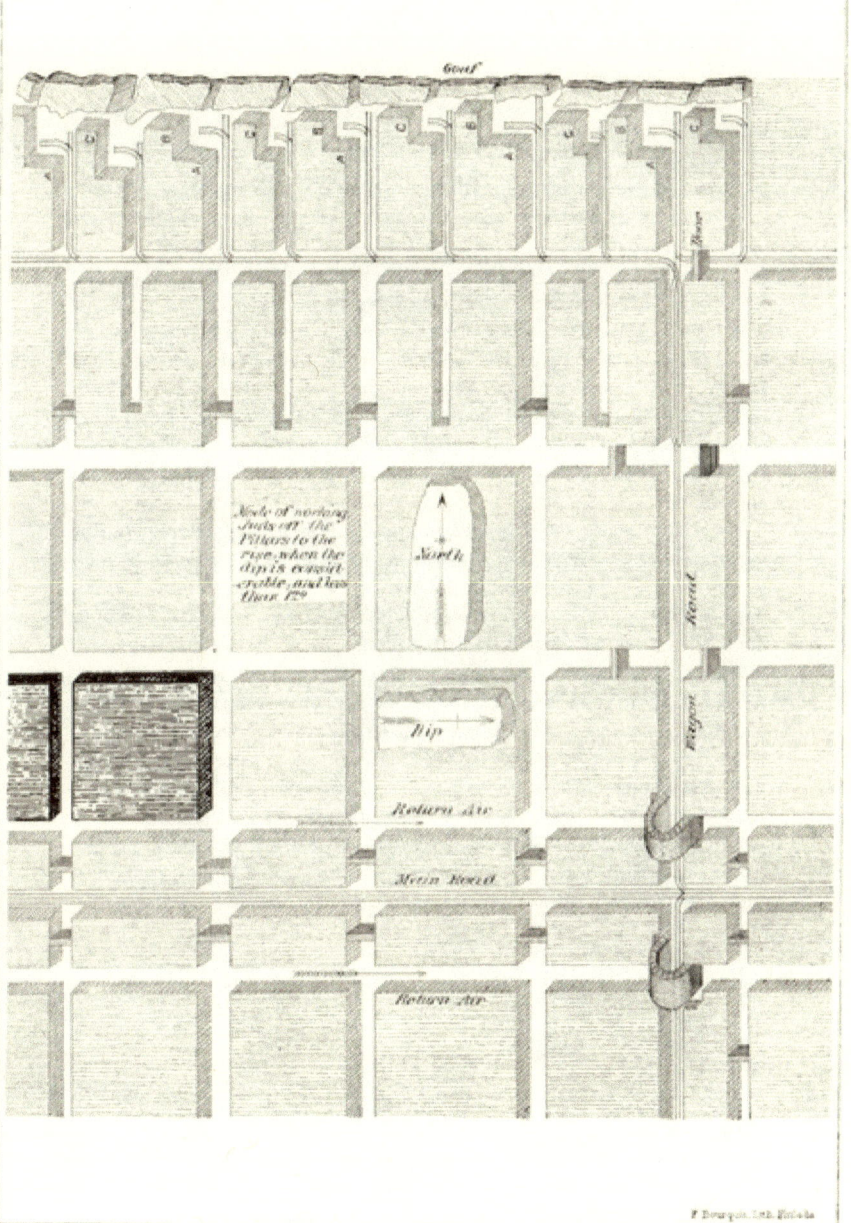

CHAPTER XIII.

DETAILS OF BROKEN WORKINGS.

"Do you not work out the pillars by any other method?"

"We vary the method of taking out the coal when the dip is too great to get the full tubs out of a dipping place. On the north side of the main drifts the coal juds are worked up hill or to the rise, as shown in Plate XIII. The pillars are split in the same manner as here on the south side; but the mode of working is only modified, as said, for the purpose of getting away with the full tubs with greater ease. In a seam of steaming coal, which does not contain so much gas, we do not split the pillars, because the pillars are left only half the width they are here. This is because the value of such coal does not deteriorate by exposure, as does a pillar of coal containing in its pores much of the light coal gases."

"Do you always work up to the boundaries of a property before the pillars are worked?"

"At some collieries they have followed up the workings in the whole coal by workings in the broken. But instances of severe explosions at some of those collieries have done much to abolish this practice. Our Plate XIV. shows how this mode is practised. You see that the goaf is little more than two pillars in the rear of the headways of the whole coal workings. The goaf in such a case is surrounded by passages which are in constant use; and we can never tell with any degree of certainty how the gases deport themselves in a goaf, which can never be ventilated in a thorough manner. It is true that the gases in a goaf may often be *drained* out of it; the heavy gases at the lowest points, and the light gases at the highest; but no dependence can be placed on the gaseous condition of the vacant spaces of a goaf, if there be any such spaces formed. And even if the fallen masses of roof have been crushed as closely as it is possible to crush any loose materials together by pressure, there must be in all goaves an aggregation of spaces capable of holding within their limits large bodies of gaseous mixtures; at times, possibly explosive in their character.

We cannot avoid this bad feature. A goaf behind us, advancing and extending itself, is like an enemy gathering behind us to cut off our retreat in time of war. This mode of working the broken was the forerunner of the *long wall* system of mining coal, which is becoming the principal mode of working coal out all over the world. In America, we hear of this long wall, or some of its modifications, being used in the bituminous coal seams of the west. In France, a peculiar mode of long wall is practised in the thick, highly inclined coal beds, and in the Staffordshire thick coal, "long wall" has been successfully applied. Instead of forming pillars as shown in Plate XIV., already referred to, in long wall, the whole breast of coal extending from *A* to *B* is carried forward, in an unbroken line, at the same time, and the coal is conveyed away from the wall-face by means of *gateways* formed in the roof, as this settles down after the advancing wall of coal. The goaf, in this instance, is traversed by those gateways, and often the main roads are built through that part of a coal mine, we have very generally termed a goaf in our maps and our reports, and our writings also."

Goaf is a term very generally understood among miners and pitmen, hence the necessity of our having made so free with it in this work, both in the plates and in the text.

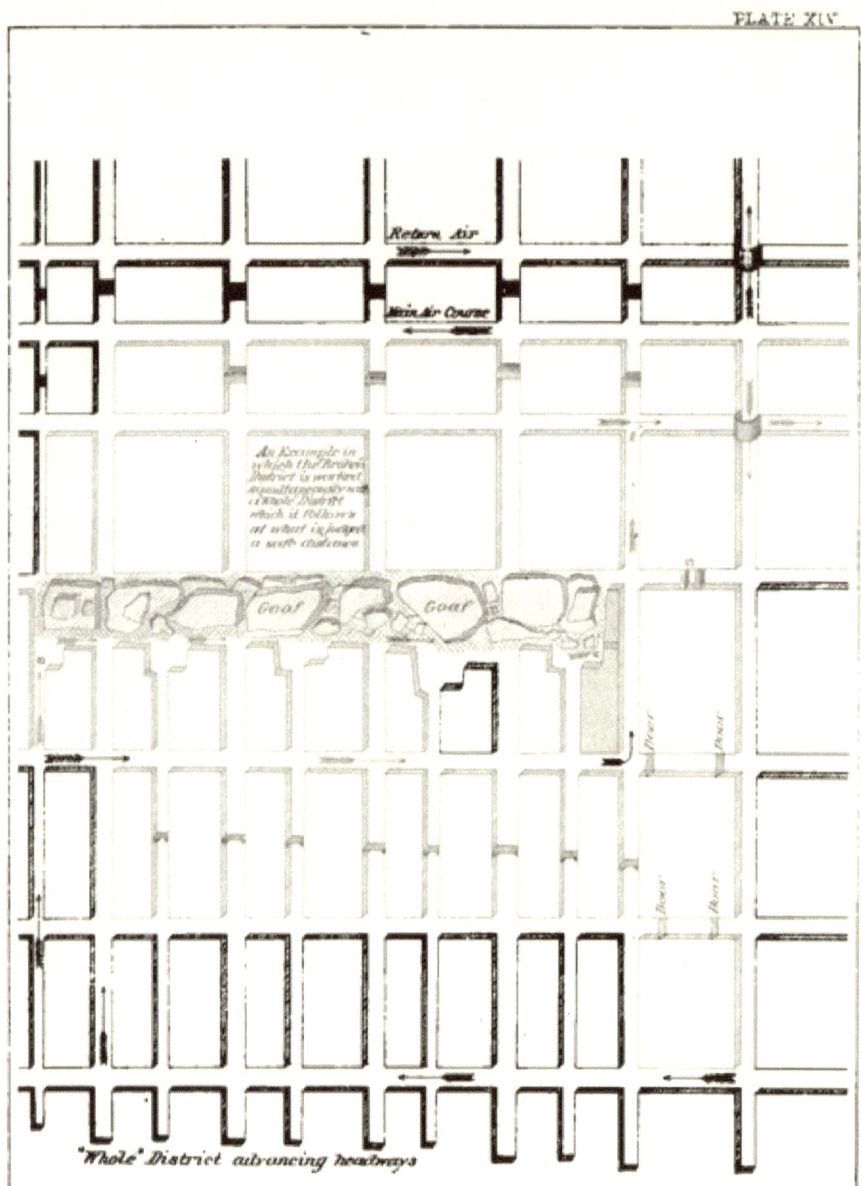

CHAPTER XIV.

REMARKS AND COMPARISONS.

THE chief merit of the plan of working shown in Plate XIV., consists in giving us a large yield of coal out of a limited area in a short space of time. It is applicable to thin seams of coal. But long wall is much to be preferred to it in every case where the roof of a seam is reasonably strong, and will stand until the props are taken out without breaking up between them.

"Is it not possible to ventilate the goaves when they are thus surrounded by passages?"

"You can only partially ventilate a goaf at best. I think a goaf charged with gas, mixed with a small proportion of air, even safer than another charged with gas and air, whose proportions are seven to ninety-three. In the neighborhoods of such goaves, nothing but the safety-lamp can be used for purposes of illumination with any degree of safety. Then, in the horse roads, and in the barrow-ways too, a creep comes on that cannot be checked, as it crushes the pillars into the floor of the seam and the floor of the seam into all the passages used for air and wagon roads; passages which must be kept open at all times, by cutting away the bottom slate as it is lifted up into the roads by the weight pressing upon the adjacent pillars."

"But, does this creeping down of the roof occur in various mines worked by different methods?"

"Certainly; we can bring the creep on in almost any coal mine by following some careless system of working; by beginning to work out pillars of coal, first on the rise, which always means the western boundaries of a property in this coal field. Then we break the strata and form our goaf on the upper side of our coal. The pressure of the roof acts with a certain amount of leverage, whose action is irresistible. The upper portion of the coal receives the full force of such action, and all passages driven in the neighborhood become completely closed up after a few months—or, in certain bad cases, weeks—after they have been opened. A large force of night shifters is therefore required to keep open the main roads. By working back from the dip with the goaf below us, or by working back in the direction of the headways, we avoid much of that scourge called the creep."

"You prefer to work by board and pillar from motives of safety?"

"In some cases only. I should use all the means in my power in a new mine to work by long wall. But there are cases where the board and pillar method would answer much better than the long wall. For instance, a body of coal left in some out of the way corner, having an area too small to make it worth the trouble of starting gateways, could be worked advantageously by board and pillar. In some of the thick coals which have good hard roofs over them a modification of board and pillar might be used. Still, I think I should not use the system in seams thicker than eight feet under the most favorable circumstances, except in mines of limited areas; and a prop eight feet in length is a very unmanageable piece of timber in a mine, except there be two men to handle it at all times. To draw such timber is not a very safe performance; to lose such timber is not economical."

"Yet in long wall you must necessarily use props in an eight feet seam?"

"Not if the seam have a division in it somewhere near the middle of it. We take out the upper portion of the seam first, and follow up with the remainder at a safe distance. The props might be less than five feet in that division of the coal which is the greatest. In very thick coals, you know very well that the coal is worked as if it were several distinct seams. A layer of coal is taken out nearest the roof, and the roof is let down on to the layer of coal under it; and as a good roof comes down in large pieces, if it does not come down *en masse*, it is just as easily managed over the second layer of coal while this is being taken out as it is over the first; and so on, till we get the roof down to the *bottom slates* on which it is to rest. Pillars of the thick coals cannot be treated in this way; therefore, if props are to be used in the working of them, they must be equal in length to the thickness of the seam. Nevertheless, there are a number of mining managers who stick to the old system of forming pillars, and waste by it lots of timber, and millions of tons of coal annually, in the thick seams of different countries. The thick coals of America suffer more from slovenly mining than do the thick seams of our European countries. But in America the mines are leased to 'operators' who pay their rents or tolls at a certain rate per ton, after it is mined and cleaned. The waste is enormous, and the land-owner is made to stand the greater part of this loss, not only of what is lost in the shape of crushed pillars, but that taken to the dirt bank, that familiar monument which is built on the grounds of every large colliery."

"But America, you know, is a new country, and the capital of their mining concerns must be built out of the natural resources of the mines. So it is natural to suppose that they will take the readiest way of getting out the coal from their mines in order to get ahead and lay by a certain stock of capital before they can be very nice in such points. Then, is it not more expensive to lay off and work large thick seams of coal by long wall than it is by the system of board and pillar as it is used in America?"

"Yes and no. Yes, because in America trained miners and *bosses* are the exceptions, and the miners are left to work so much according to their own ideas, that it would not at many places be prudent to set them to work on the long wall plan. Both miners and their bosses would have to be picked carefully from among the people working in the mines. Until the system of long wall gets a good start in America, so as to get the miners practised to work it, it would occasion some expense. In the West there are good opportunities to begin long wall, because the seams lie very favorably—slightly inclining—and are not so thick as they are in the anthracite coal fields. If men are employed to work on the system of long wall who have not been trained to it for a short time at least, the system will be more expensive to work than board and pillar, which all miners seem to understand sufficiently to work to tolerable advantage, were their labors well directed by competent mining managers. It would not be more expensive to lay off and work long wall if a good corps of *gateway men* could be obtained, with a good foreman, having *practical experience* at such work."

"Then you prefer taking out pillars by working back in the direction of the levels, beginning at the dip or with those levels driven on the lowest part of the boundary?"

"Certainly; and never on the highest levels unless circumstances are extremely favorable, or for some reasons it be imperative to do so. If the dip is above $25°$ it will not be prudent to begin on the dip, because the upper portion of the coal will be in a manner undermined by the lower portions."

"In the pillars of the south side, as per Plate X., the juds are worked off to the right and left of the headway splitting the pillar; do you ever depart from this plan to any extent?"

"Where the pillars are thick and the boards are choked up we split the pillars, in the manner shown on Plate XIII., and if the dip is great we work only that coal on the rise of the splitting headway. We get the other coal left below the splitting headway by shearing off a portion of coal along the high side of the old board at the points designated by *A* in the same Plate. The juds *B* and *C* may thus be worked off simultaneously. We save the expense of helpers to get out the full tubs, and the men fill down hill instead of up hill, which is a matter of great importance in the performance of a day's work, where so much coal must be handled by the men. But as far as working the coal is concerned, the system is not at all changed."

"As it is a considerable length of time before the *broken* of an extensive coal mine can be worked, when you adhere to the plan of working backwards, in what condition do you generally find the pillars and the excavations surrounding them?"

"Very often they are closed up. Some of the boards are closed by falls of roof and others are closed by an upheaval of the floor. From this latter fact we have applied the term *metal ridge* or *rig* to the material found in all such closed-up places. In working the *broken* in such mines, we are obliged to tunnel through the metal rigs; and then we either split the pillars, if they are of a sufficient thickness, or skirt along the old boards by taking off a strip of coal. (See Plate XV.*) When such places are worked, the *metal rigs* are a source of trouble and expense; and this expense is greatest where the pillars have been left the smallest in area. In some of the oldest mines much of the coal left in pillars over a hundred years ago has been mined; and at the present time many mines are working out the old pillars. At the Lambton and Washington and other collieries, works are in progress getting out the pillars of the old mines. We find such pillars to be as a rule thin; but there is quite enough coal in them to well repay the trouble and expense of working them out. Plate XV. shows the manner of working through the *metal rigs* of those mines when the bottom is lifted."

* Plate XV. shows how the drifts of an old mine are often found closed up by the *creeping* in of the bottom slate. It shows also how these old mines are operated the second time over by driving through those *metal ridges* formed in the old excavations.

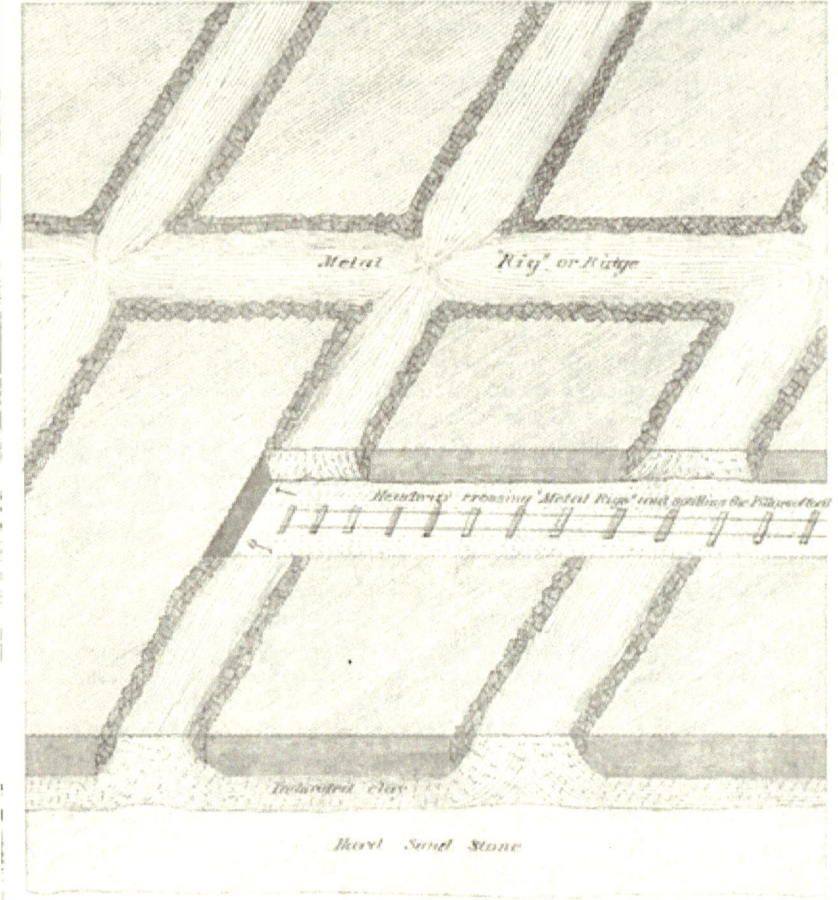

Working Coal by Crossing the
Metal Ridges of the old
Board Rooms

Isometrical Section and plan

"As you work forward toward the boundaries of your mines, we are to understand that the board-rooms always close up by the upheaval of the bottom or by the falling in of the roof?"

"In many cases this is so; but there are instances of places standing securely for nearly a century. Much depends on system. For instance, as we work boards to make coal as well as to block out the pillars, we should work those boards narrowest which are the nearest to the main roads and the furthest from the boundaries, because those boards are the first within a district to be opened; and for this reason the pillars they surround stand the longest under the weight of the roof; if this should lose its *lateral* support there would be a crushing down of the roof on those pillars in the vicinity. We do all we can to avoid the pressure being brought on to those boards nearest the main roads which run through the central part of the property (royalty we term it in our phraseology), and we leave the pillars as large as circumstances will allow us to do to prevent, as much as we can, the creeping down of the roof. But a roof which contains abundant supplies of the bisulphate of iron (pyrites) chips off and falls into and closes up the board-rooms in spite of all we can do to prevent such an occurrence."

CHAPTER XV.

RE-WORKING OF OLD MINES, "METAL RIGS," AND OLD COAL PILLARS.

"IN Plate XV., showing the mode of opening work through old board-rooms closed up by ridges lifted from the bottom slate, the pillars are small; such as they were made a century ago, in such mines as Ravensworth, Fatfield, Lampton, Washington, Black Fell, and others opened early in the eighteenth century. These are the oldest of our mines worked at the present time, and the matter found in the old board-rooms of such mines is very compact, whether it is from falls of roof, or upheavals of the floor. In case of the ridges being thrown up from the bottom slate, it presses so tightly against the roof as to form a very excellent means of support, being in fact nothing less as a whole than a network of pillars, standing as regularly and systematically as the places were driven in which they have formed; and we find the old workings of those mines referred to driven very regularly; that is, the boards are found very nearly at equal distances apart, and they approximate very closely to parallel lines. When we have *falls* in the old board-rooms, the material of which has not become packed closely by a long-continued action of pressure, we use timber to keep the loose stones remaining above the height we require for a passage from coming down. It requires careful men to timber such places. The greatest trouble we have to contend with in them seems to be in *getting* in the first set of timber after we leave the coal, and also the securing of the last set of timber on reaching the coal on the opposite side of the old board-room, or other excavation. This is owing to that line above the edge of the coal pillar being comparatively straight, and this facilitates the starting off of the slates and rocks. In the more central part of the old boards, the rocks or slates are interlocked with each other in such a manner as to support themselves laterally. But, if by carelessness, some of the rocks get a start near to one side or the other of an old board-room, it is a very difficult matter to avoid a 'runaway' of the balance. This is the more liable, if the 'stuff' in the board-room is of a friable nature, or lies in small cubical pieces. I have seen a hole in the timbering, not larger than your head, empty all the rubbish off the 'forepoling' of the timber work, in just the same manner as if it were a body of gravel. Such loose material will run off like a mass of shingle or pebbles. But generally those 'metal rigs' are managed very well, whether they be composed of matter from the bottom or from the top of the seam. The reason of this is because such work is done by the 'shifters' under the supervision of a 'maister shifter,' and this, with the bulk of the repairing of the working roads—not including the returning air roads—is performed during the night. But on certain pressing occasions, if a hewer strike into a metal rig during the early part of his shift, he is employed to cut through it, and paid according to the time he is at work in such a place, together with an allowance called 'consideration money,' which is supposed to compensate for any loss of time incurred in such an event. The deputies regulate such matters, being always on the ground, and more competent to do justice, both to the owners of the colliery and to the men who perform this or any other necessary work for the benefit of the colliery."

"How far can you drive your headways in a given time in the *whole*, when the coal is of ordinary hardness; such coal, for instance, which has not been drained of its elastic gases by approximate workings?"

"We can drive the headways, opening the coal panels at the rate of half a mile a year, without the least hurry."

"Then, to work off half a mile of panel it will require four times this length of time at least?"

"It will require five years to open and work off a panel of coal half a mile in length, if the seam is a thin one, and one which may be worked off at the first working. In the Lancashire coal mines, and in some others, where the dip is above 15°, the panels of coal are not laid off and divided into pillars, as here described until the main headways of each district or panel have reached the mine's limits. Occasionally, however, the panels of coal lying between any two pairs of headways are intersected by places driven in pairs to the drifts on the rise. In such works, all the levels or endways are driven in pairs, and in some cases triplets, at a distance apart to suit the ideas of the manager, and those pairs or sets of drifts are connected together at short intervals of may be 300 feet to shorten the air course and carry the air forward. As soon as the boundaries of a mine are reached by this plan, operations of working the coal panels out may be commenced at once by blocking them out into pillars. This is done by commencing to cut headings or endings backwards, and on making the board-room connections at equal distances apart we have the board and pillar system before us again; but it is so modified that instead of advancing we get our coal by retreating from the boundaries. By such a practice, a panel of coal half a mile long by one hundred yards, or, in some cases, two hundred yards in width, may be worked out in five to seven years. We are premising the thickness of the coal to be from three to five feet, the thickest beds requiring the longest periods to be worked out. But where the dip is considerable, it is seldom that the distance between the pairs of headways exceeds one hundred yards. In the thick highly inclining coal beds of France, this system of working was much practised until quite recently. In that country, under favorable circumstances, and with good management, very good results have been obtained by its practice; but the mined coal was often covered by the falling rocks, so much so that it has almost been abandoned. In Lancashire this modification of board and pillar is still in general use. For the mining of coal lying in seams less than seven feet in thickness, its main objection consists in the deferred production, owing to the length of time required to advance the levels to the mines' extremities. By laying off the shafts, so that the distance to the boundaries will be much less on one side than on the other, the full produce of a mine may be sent out from the short side, whose drifts have been driven home, and the broken commenced, even while the headways are still advancing on the side farthest from the boundaries. In many mines lying on a low inclination where the creep may be kept off for a great length of time, it may be more prudent to mine out the coal of a panel half a mile in length, within eight to ten years, instead of within five to seven. In other words, the panel of coal worked out as in Lancashire, requiring five years, would not be worked out in the Newcastle Coal-Field—as a general rule—in less time than eight years, because the pillars being cut out while advancing, so large a yield of coal is obtained from the board-rooms and headings as to render the pushing forward to the mine's limits, only a matter of secondary importance."

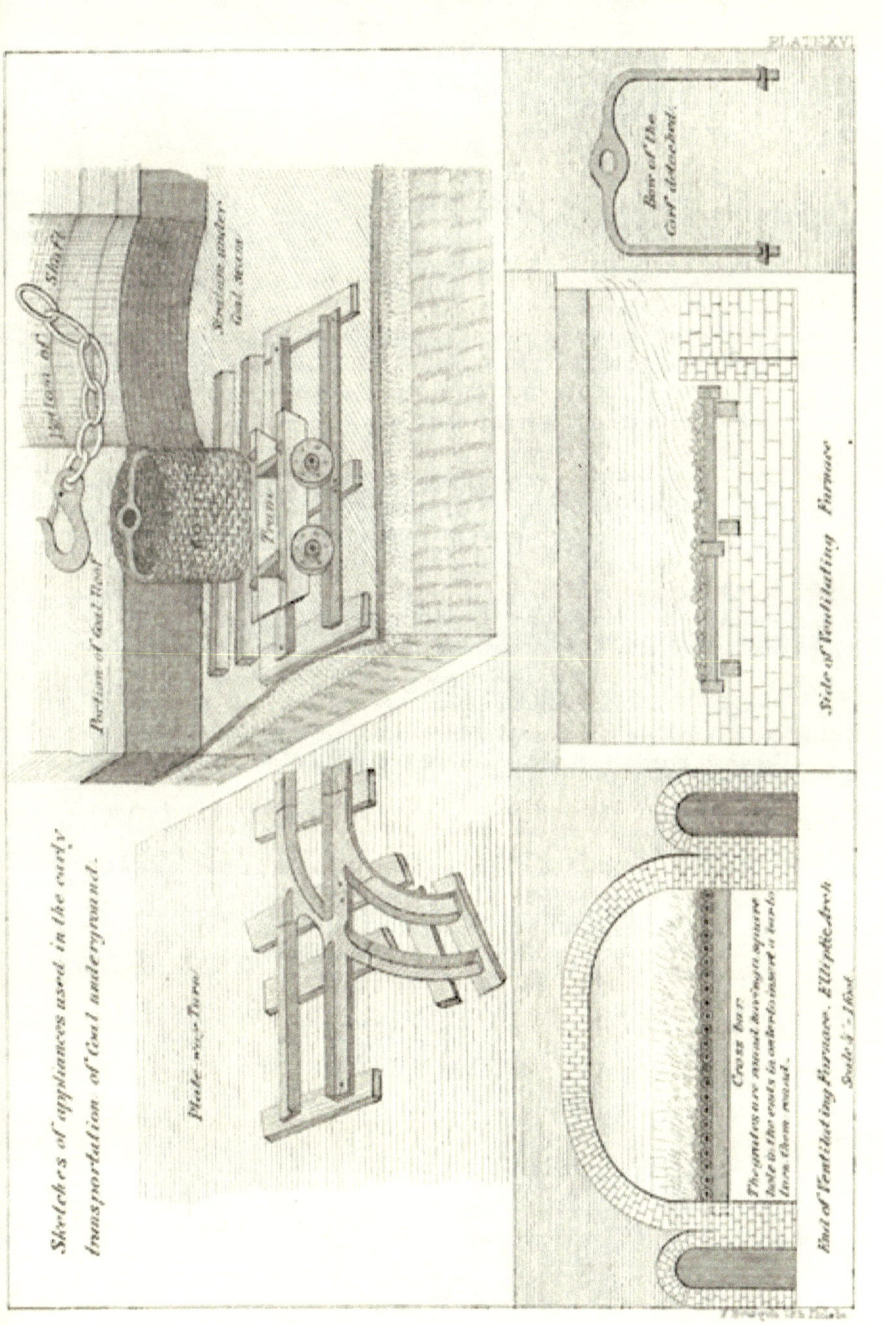

CHAPTER XVI.

GENERAL REMARKS.

WE are now in the OVERMAN'S CABIN (Plate IX.), of a well-conducted coal mine; and, with the reader's permission, shall detail the results of a familiar chat concerning the men and the mines of the Canny Newcastle Coal Field. Here the foundation of railways was laid. The willow basket first used by the old women who carried coal home to kindle fires and boil their teakettles, grew into the *corf* made of hazel sticks. The handle of this became a *bow* of iron. Then it was too heavy for a strong man to carry; so he placed it on a frame running on small wheels which he called a tram, and pushed it over the smooth floor of the coal seam. The corf grew larger, as the man or strong boy who pushed it became more expert, and this boy received the name of *putter*. When a wooden rail, with its flange or guide, suggested itself, the corf grew so large as to hold, in some instances, five hundred weight. Then, as the roads were to run into places started off the mother-gates and main headways at right angles to them, the TURN was invented. But, while the wooden rails lasted, the *turn* never was very satisfactory as an adjunct; the putter seldom passed a turn without jumping the track and bestowing on it his most emphatic curses. But the time came, which brought tram plates made of cast iron. Then a skilful pattern-maker, or *joiner*, got up patterns at the request of a clever *viewer*, and the cast-iron turn, as it is shown in the Plate XVI.,*

---

* In Plate XVI. we find some of the older mining appliances. The hook we see, when in the hands of an expert onsetter, could be thrown into the eye of the corf's bow every time without fail. The details show us the corf, tram, tram plates, and turn. The pointed projections shown on the tram were necessary to hold the corf in its position. To enter the turn with a tram, the putter knew well how to twist his tram to suit his particular gnit.

When the tram and corf were in use, little boys were employed to brush the tram plates several times a day. The dust and fine coal would trickle through between the hazel sticks of which the corf was made, and cover the road, which proved to be a source of danger, as well as an inconvenience of no slight magnitude; of danger, because of the dust mixing up with the air, to contribute to the force of was contributed as an advanced step towards the formation of the modern railway. The putter was delighted with the new cast-iron road and the tram, and being himself the locomotive engine of that period, he was in the economy of mining an important item. Of course, the tramway had crept out of the pit, and it was laid on a grand scale on the surface; and wagons, having cast-iron wheels, and carrying a *cauldron* of coal (about 52 cwt.) ran over the tramway, which took the respectable name of railway, when it came out of that dingy coal pit. But the difference between the railways of the surface, which ran down to the rivers Tyne and Wear, was that the flange of the rail was transferred to the wheel of the wagon! From this point of the railway's history, improvements came so thick and fast then as to take the breath out of the point of a man's pen now, and, in his confusion, he begs to point at the magnificent railway system of to-day!

Many a bright fellow has contributed brilliant ideas before all this railway magnificence has been attained. We find enough to say of such things as these as we sit chatting with the overman in his cabin.

Of the sad times caused by the MISFORTUNES of the past, he speaks with feeling and intelligence. "We know more of gas, now," he says. "Our fathers did not know what was the weight of a pound of air! They had no scales like these hanging on the walls of my cabin to weigh and balance the air at any moment." He signified the barometer, which is to be observed as often as possible, and its movements noted as often as they are manifested. "They knew nothing of the expansion of air and gases by the action of heat, nor could they tell exactly how it was that the escape of gas was the most abundant during a storm. The safety lamp is only a modern invention; they had not the benefit of it at the old collieries. They had their *flint mill*, giving them light from the shower of sparks it emitted; but they were very clever with their candles; they could work by candle-light, until the gas

---

an explosive mixture; of inconvenience, inasmuch as it retarded the passage of the tram wheels. The original tram wheels were broad on their bearing surfaces; but experience brought out a wheel with a sharp edge, which ran easier over the dusty tram plate.

An old ventilating furnace is shown in this plate. To keep the air passing over the fire as near to the heat as possible, an elliptic arch was used. The more the heat given out to the air-current, the greater is its ascensional force or draught in the upcast shaft.

in the air was within a shade of the *firing point*. But, with all their skill in the use of the naked light, we have recorded, in the past history of our mines, the terrible explosions which occurred at Newcastle, Washington, Fatfield, Springwell, and other of the old pits. The deputies and overmen could not be at all points watching the gas, as it strengthened in the air by its accumulation. I have heard tell of the manner of lighting up some of the first furnaces used for the purposes of ventilation. What would you think of this plan of lighting up a furnace by running a large link of red-hot iron down the furnace shaft on to a mass of combustibles placed on the grates of that furnace? My grandfather tells me this was done at one of the Fatfield pits, when the furnace was quite a new institution in the coal field, and when the ventilation of the workings was effected entirely by the means of coursing the air all over the works in one single body, causing this body of air to travel through all the passages in succession. Of course, with such ventilation the workings of a mine could not be extended beyond certain limits. It was necessary to sink very many shafts, in order to assist the ventilation. Each district would require a pair of shafts to ventilate it properly, an upcast and a downcast. And there were doors at the end of every headway's course; and a door in a main road is always an eyesore and source of danger as well. As you see by reference to Plate II., by the use of our air-crossings we dispense with doors entirely in the main roads. For the sake of convenience, however, we use an occasional door in the barrow-way at this time. But, before jumping over a couple of centuries, we must say that the idea of coursing the air around our mines occurred to Spedding, and this system of ventilation was considered in his time a wonderful invention, and we must so consider it now. It was the origin of our present system. When John Buddle took the half of the main current of air descending a shaft, and conducted it into the workings of one side, and allowed the other half to flow into the workings of the other side of the shaft and ventilate them, there were grave remarks made concerning the wisdom of such a plan. But, nevertheless, day after day succeeded each other, and the plan worked well. At the bottom of the downcast the air separated into two distinct currents, and each current went in and ventilated a district, and then they both came out through the return courses, bringing off the gases, and meeting again at the bottom of the upcast, up which they went harmoniously together. I maintain that each current was worked according to the ideas of

Spedding. We can now extend our districts, and multiply these currents (*splits*) and provide every district in a mine with a distinct current of fresh air, and say that we are indebted for our fundamental principles to Spedding and Buddle."

"But the officers of your coal mines have kept well apace with the recent applications of the arts and sciences as applied to coal mining, and in some important cases have taken the lead in inventions which have not confined their merits to mining, but have become of great general use in the civilization of the world. I can never stand in the engine-room of one of our splendid Hudson River steamers without thinking of the similarity existing between the engines of those steamers and the engines you use to wind the coal out of your shafts. I hear the gong of the steamer giving signals to the engineer on duty, and I think of the 'rapper' of your older pit heaps, which informs your engine-man to 'draw away,' 'bend up,' 'lower down,' and the numerous other signals you communicate with the rapper, which we call a gong, being in principle one and the same instrument. Are not the mining engineers of to-day educated in technical schools?"

"The technical school of our mining engineer is and has been and is likely to continue to be the mine itself; and every mining engineer having an apprentice in charge is his teacher. Coming to the engineer's office with a good scientific education, our young viewer has to begin at once to get his practical experience, which is always that part of any one's education that comes the most slowly; but still he reads under the direction of the engineer and continues to complete his education. We who hold the subordinate offices in a coal mine are, to a certain extent, benefited by the education of our superiors. We become, by our daily intercourse with them, invested with their ideas. We read the works they write, and as we consult daily concerning the condition of our mines, we must agree on all important points concerned in the working of them. In fact you may take the force of a coal mine of our district, and you will find that it is a united force chained together like the links binding the problems of Euclid. The trapper is taught to open and close his door. A link further on our trapper becomes a driver, and he is started at his new post with one of the *canniest* horses we have. His next step advances him to the position of a 'galloway driver' or 'pony putter,' when he begins life on his own account; because he then starts work by the score or by contract. The spirit in the boy is brought out, and in nine cases out of ten, he becomes eager to earn as much money as he can. As a driver of a horse, the putters would awaken him out of any slow, sleepy movement should he come *in bye* too tardily. The wagon-way man would roar out, 'Drive away, lad! Don't fall asleep!' etc. But the moment he gets his initiation into the life of the putter, whether he have a pony and sixteen cents a score first rank, or whether he *puts* his coal by the strength of his arm as a hand putter at twenty-eight cents a score first rank, he needs no one to look after him. As a pony putter he will earn sixty to seventy cents per day. As a hand putter he will often earn one dollar and a half per day; more money, in fact, than a miner. We do not need to 'hurry up' our putter lads; stimulated by their running in opposition to each other, the lads do all they can to exceed and excel each other; and it is astonishing to see how bitterly they will contest with each other over a disputed wagon when two of them come to a station by different ways at the same time in a 'dead heat.' They will quarrel and get to blows, even though they be the best of companions when out of the pit. But this spirit is the life of our mines, and as it goes with the lads into the face when they begin to cut coal, we need only know that everybody is in bye at 'calling course' to know that a full day's work will be sent to bank. The entire pit may be a technical school to any one who wishes to learn in it."

"You mention two terms which need explanation: the first rank of the putter, which has some reference to his price; and this 'calling course,' which brings him to his work at the proper time."

"First rank means that when the average distance of the men's places into which the putter goes for his coal is eighty yards and less than one hundred, he receives twenty-eight cents for a score of tubs brought out to the station. At one hundred yards we give the putter his second rank, and an additional price of a penny (two cents). Every twenty yards advance is an additional rank, and adds another penny (two cents) to the price of the putter's score. The pony putter's first rank, which is less than one hundred yards, brings him but eight pence or sixteen cents a score; but his other ranks, of twenty yards each, bring him an English penny each, in addition."

"And *your calling course* is an institution different from what it is found in other coal districts!"

"It is a regulation which is of the utmost importance to us and to the men employed at our collieries. One of the great advantages we possess over other coal districts is, that colliery owners own the houses the men live in. No rent is charged except a small nominal sum, kept off at pay day, which amounts to three pence (six cents) per week. Coal is laid down at our men's doors without charge. They pay no taxes or church rates. The colliery houses are in rows, and each colliery has a village of its own, and each house has its potato patch, and many have flower and fruit gardens as well. The houses being thus together, the *calling course* is very easily effected by some of our old miners who have good lungs and hardy constitutions. The caller is generally furnished with a pit pony, and his first duty is to get around among the houses to call the *fore-shift* men. This he does at about two o'clock A. M., beginning with the deputies, who get the start of the men by one-quarter to one-half of an hour. As the men change their shifts each week, the caller must bear in mind this fact and call only the men who go to work in the fore-shift. He is, therefore, required to bear in mind the names of the men, who change from the backshift to the fore-shift every alternate week. His next duty is to *waken up the lads*, which duty he performs two hours later. As the lads include all those who are employed in the transportation department of the pit and who get up at the same time all the year round, this task is more easy and is attended with less confusion and perplexity. Among the lads of the caller's list you find the names of the *banksmen, screensmen*, the *onsetters, waggon-way men, drivers, trappers, putters*, and the *brakesmen* (enginemen). The caller is the village time-piece, and every one lies in his sound sleep until he calls out or lets the rat-tat-tat of his stick fall upon the door. He never leaves a door until he receives an answer, and if this comes so tardily as to cause an unusual loss of time, the caller remembers the fact the morning succeeding, which is manifested by the increased vigor of his blows on the sluggard's door. The caller has no time to spend at the door of any one, and he does not forget to let those know it who have given him any extra trouble, and who have been the means of causing the loss of a few seconds of his time. I tell you, the wind may blow, the rain pelt, and the snow drift, the caller must get through the whole of it and accomplish his task every morning at the same time all the year round, Sundays and sometimes *pay Saturdays* excepted. The shaggy Shetland pony, invariably his faithful companion, seems to know as well as his master at which of the doors he must stop, and he leaves with his best gait and gets to another door as soon as the satisfaction required has been signalled from within. The consequence of such an arrangement is that the 'lads' come to the pit at very nearly the same time. The pit heap, silent and deserted a minute before *calling course*, becomes a short time afterwards a scene of great activity. The lads flock in and descend the shaft as fast as the cages can run them to the bottom, and at the moment the last cage full of lads gets to the bottom the pit begins *coal work*, and seldom do the ropes remain idle in the shaft from this minute until it is time to quit work."

"You have a spare hoisting shaft you use to draw the water out of the mines in those large tanks; is this used for any other purpose?"

"During the daytime the tanks are replaced by cages, and we draw coal through this back-shaft. The cages, to be handy, stand on trams and are readily hooked on to the ropes after the tanks have also been run out of the shaft on trams and disconnected. The trams run on rails laid into the shaft for the purpose. These rails work on hinges and form gates when the cages or tanks are running, and they are adjusted in a few seconds. They form also the guides of the cage as this is taken to a higher level to land the coal tubs."

"But this is not the question we wished to ask—if we are not getting too far into the mechanical department—is not the changing of the men at half work a cause of delays in the winding of the coal, and would it not be preferable to have a shaft accessible at all times for the purpose of the lowering down and the drawing up of men and for the sending of stores into the mines?"

"Shafts have often been provided for these purposes at several of our collieries; but as soon as the coals have come out of the mine in such abundance as to be beyond the capacity of their hoisting facilities, those spare shafts generally have been changed into shafts for the winding of coal. They may serve the purpose of avenues for the supply of timber and other material after quitting time as well as for the drainage of the mine; but to see them idle seven-eighths of a day, with another shaft crowded with business, is more it seems than can be allowed. A hoisting engine and a shaft in this coal field, with shaft fixtures to match, are a very costly investment, and to keep them idle, or nearly so, would not be a prudent measure to pursue. As our ropes can change each other with men on once a minute, we scarcely feel any delay in the lowering and drawing of men, while these are changing their shifts at half work."

For the present we shall bid adieu to the board and pillar system of working coal, by calling the attention of the reader to Plate XVII.,* which shows the interior of a board-room, the manner of branching off the road into it, and the mode of propping under a slate or shaley roof. The castings of the turn-off consist in the points which are fixed and open, the crossing and the sweeps are all made of cast iron. As shown, they are gauged when they are laid down so as to fit each other, and the out sweeps being struck off from the long radius on the outside, are the longest, and are of an easier curve than those of the inside rail.

The *board* work is the safest and more to the miner's taste than any other; and for this reason it is a mode of working which finds favor in new districts where miners are scarce and where coal is abundant; but if care is not taken in the pursuance of it, much coal may be wasted by being left in the mines in the shape of pillars, made inaccessible to a second mining; or by the working out of those pillars, as in the *broken* workings.

In England, where this system is practised, the coal is leased by the acre, and if any part of the coal is wasted in the ground, it is paid for notwithstanding at the same rate as it would be if it were taken out. In the United States, where the coal is paid for at the rate of so much per ton, the coal is mined in the most reckless manner, and for this reason much coal is wasted in the mines by the caving in of the roof. As the next section treats of those mines of thick coal which lie on an inclination of over 40°, it is hoped that after its perusal any one may see how near to an impossibility it will be to mine out completely those pillars, when they are left standing at each side of a large excavation, on the floors of the seams which are so highly inclined.

* Plate XVII. shows one of those single *turns* which are used for the *barrow-way*, and how it is made of three pieces of cast iron, besides the sweeps or the curved pieces which are necessary to complete the right angle. It is laid in the headway's course and branches into the board-room newly opened from the headway.

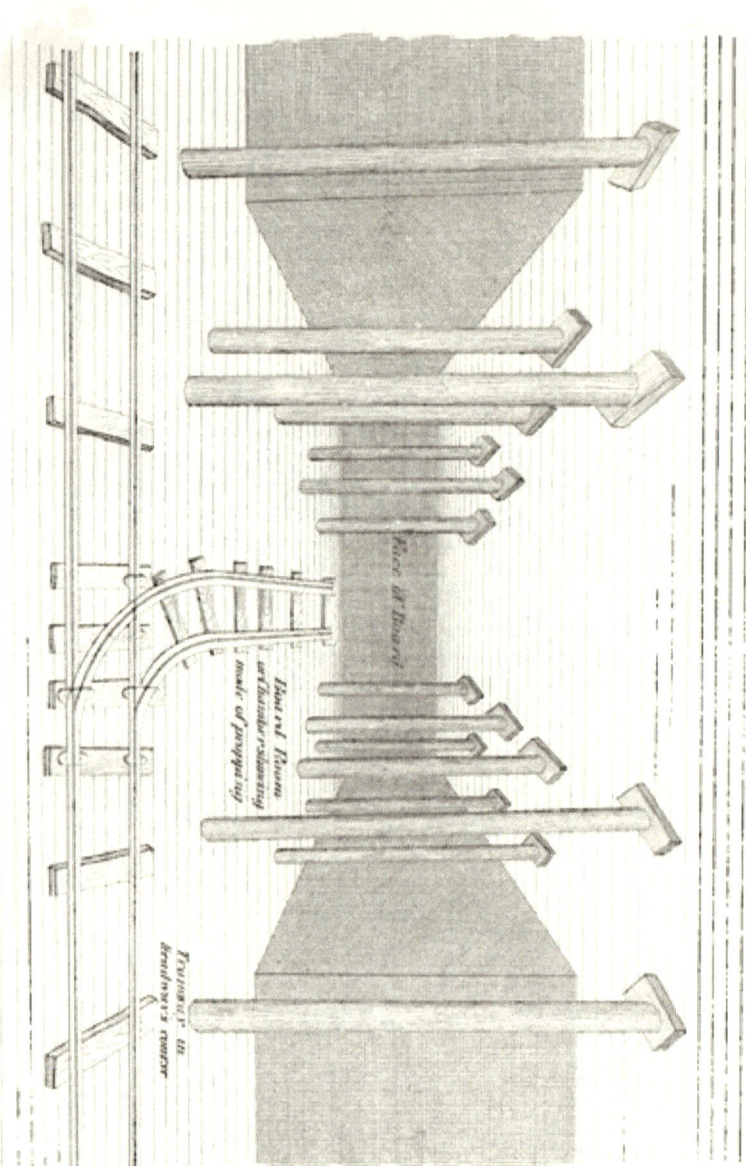

SECTION III.

HOW COAL IS TAKEN FROM THE HIGHLY INCLINING COAL VEINS OF THE UNITED STATES.

CHAPTER XVII.

TOPOGRAPHICAL FEATURES—CHARACTERISTICS OF THE MINES.

In one of the gaps through a coal mountain are situated the outlets of an extensive coal mine. The mountain is within the limits of Schuylkill County, Pennsylvania. The coal strata are broken entirely through, and the wide breach runs down through the great bed of conglomerate on which the coal measures rest, as do those of England on the mountain limestone. But our breach continues to a great depth into the red shales below the conglomerate. The masses of coal strata thus broken apart have been thrust away from each other by some huge wedge of rock, pressing upward on the red shales, receiving its motion from the action of some subterranean force of great magnitude. The line of division is nearly east and west and approximating a parallel of the main faces of cleavage peculiar to the locality. But here there are two great mountains made instead of the one above mentioned, while in the breach a valley of a pleasing topographical aspect, with its modulating surfaces, stretches out between them in perfect harmony with both. That mountain to the south is the Broad Mountain, whose ample plateaux have been a sad puzzle to the intrepid explorer, and which have often been made the subject of some greedy speculator, too timid himself to thrust a drill down into their mysterious folds of stratification, and too crafty to compromise himself with any one more plucky and more able to fathom its secrets.

This mountain, to the north of the valley in the red shale, is the Mahanoy, and the gap mentioned at the commencement of this chapter cuts it clear through; and here the Mahanoy Creek finds its way from the valleys lying to the north, as its waters run off to join those of the Susquehanna nearly a hundred miles to the southwest.

Within the flanks of the Mahanoy Mountain, the coal strata are found dipping north, and lie on a high angle of inclination. They run down to a great depth below the level of the creek in the gap. But they also run upwards to a great height, tipping up the outcrops of coal seams literally into the clouds. The ridge of the mountain is formed of that hard enduring rock which covers the famous Buck Mountain coal.

But the Buck Mountain vein, important as it is in extent and quality, is not the chief vein cut open by the gaps. There is that Mammoth Vein, whose jet black diamonds have a world-wide reputation, whose thickness approximates twenty-six feet, the products of which are so excellent in quality as to have acquired such high-sounding titles as " King of the Lehighs," " Kohi-noor," and the rest.

The outcrop of this vein is on the flank, and its place in the strata is about two hundred feet farther away from the conglomerate than the Buck Mountain Vein.

The drifts of the coal mine referred to are driven above the level of the creek, and they pierce the Mammoth Vein. Confining ourselves to the operations of this Mammoth Vein, we present an example of working a thick coal seam, lying on a high degree of inclination; and, to better illustrate all operations of importance, we adopt the second person to treat the subject, and to begin, we introduce you to the notice of the reader.

In Plate XVIII.* you find a mining representative sitting on a pile of mine timber, smoking his pipe, and gazing after " the trip" that has just emerged from the drift mouth, and is on its way to the breaker.

He is a burly Yorkshireman, with a round bloated face, with whose amplitude you fancy John Barleycorn has had much to do.

" Can you give me employment as a miner ?" you ask.

---

* Plate XVIII. presents us with a view in the gap or ravine cutting through a mountain. Could we remove the surface soil, we would have the coal strata before us as they alternate with each other, like the leaves of a book, and as closely associated together as only dame Nature knows how to connect and harmonize her works. When the ridges of these mountains are sharp, and their sides and slopes steep, the strata generally incline at a high angle. Fortunately, this places the coal beds within easy reach of the miner, and as they are elevated in part to various altitudes, those portions above the water levels of the neighboring creeks are easily drained. For reasons already shown, the timber is rapidly disappearing from the surfaces of the mountains. The part of the sketch which contains the most of interest to the miner is the trip, which represents the manner of transporting the mine's produce, the outlet for which is seen to the right in the form of the water level gangway.

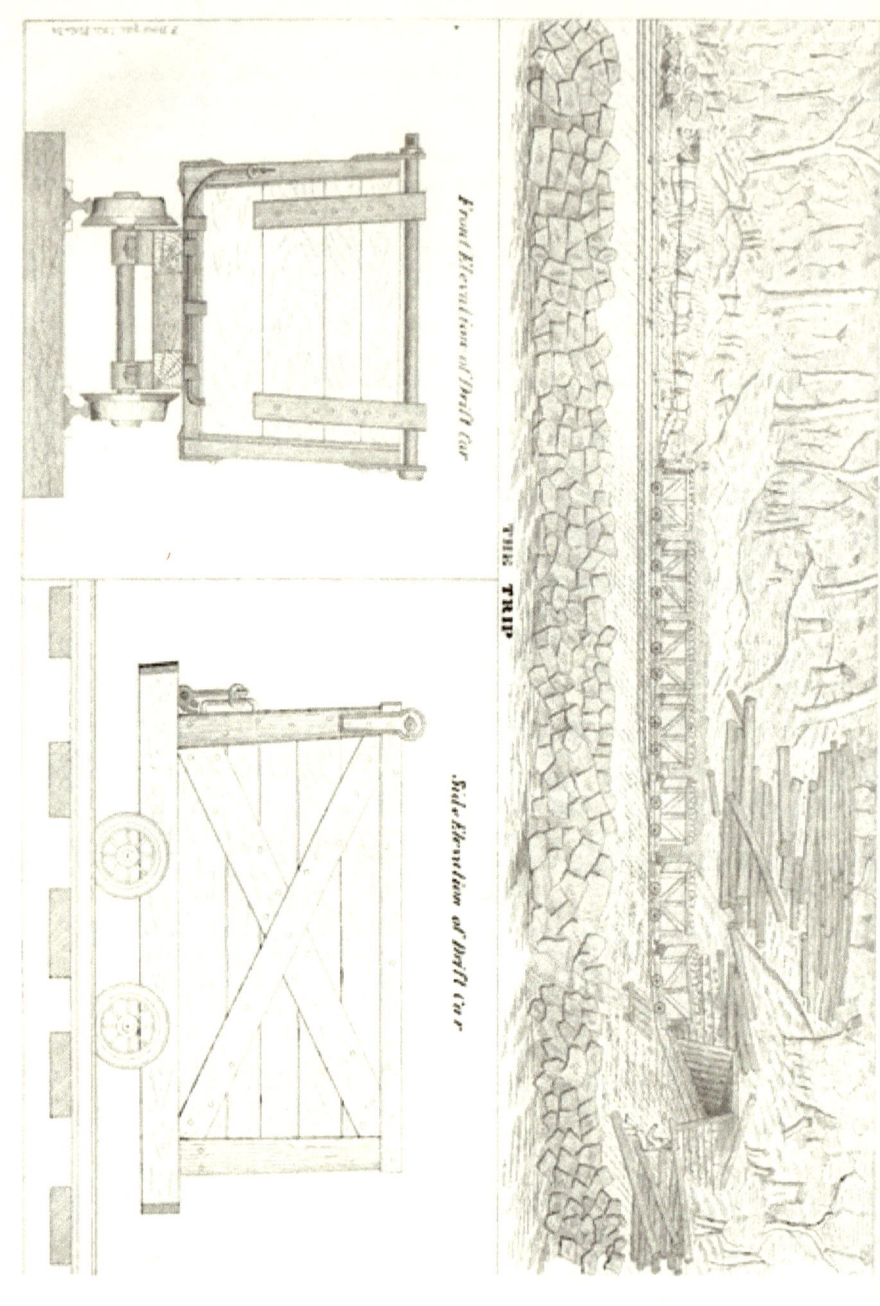

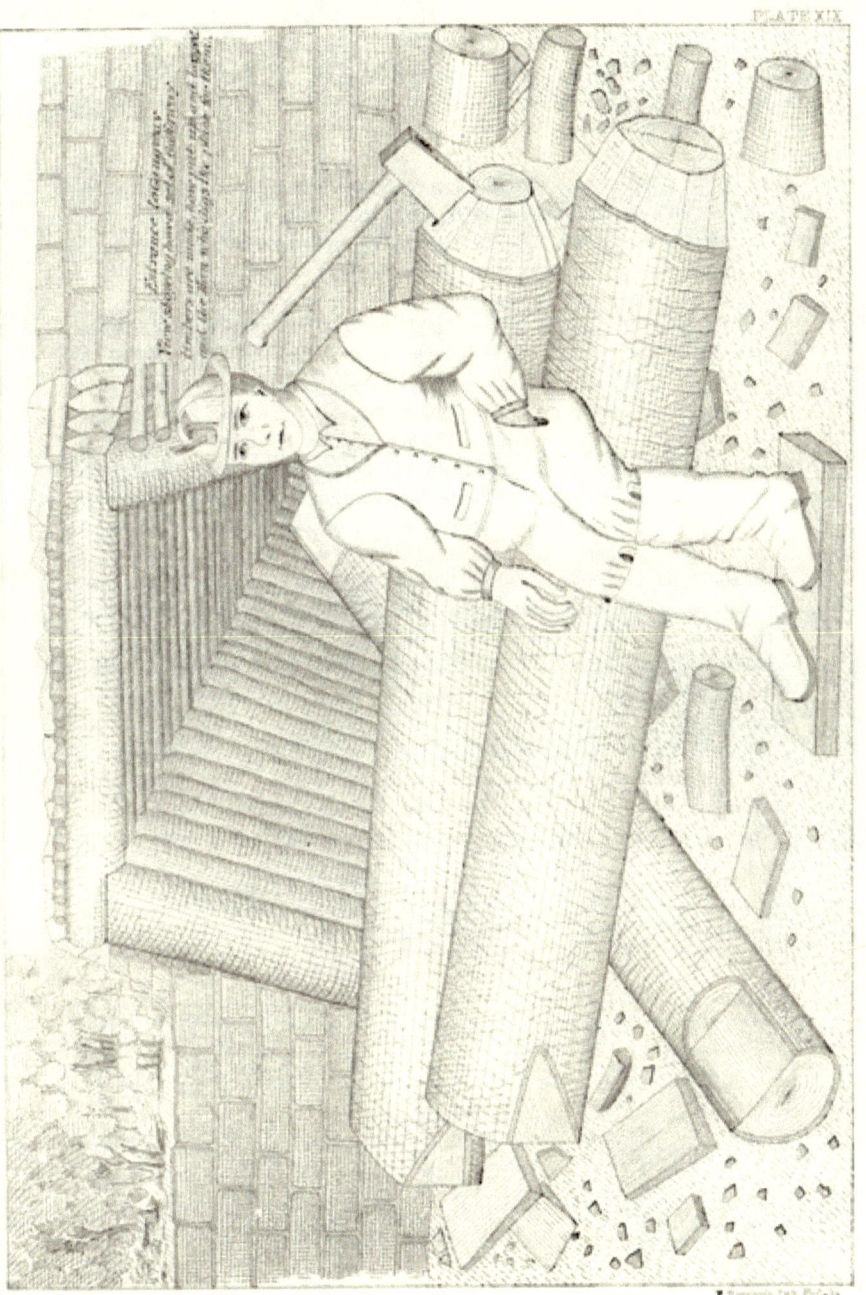

The effect on the stout gentleman is electric. The pipe comes out of his mouth, and the reflections of his mind, whatever they may have been, are recalled, and you become an object of his scrutiny; and as the following will show, his criticisms also.

"Tha wants work as a miner, does 't' better get a job as a machine agent! Thee art' n't a miner! Where has 't' worked last?"

"I have worked last at the Trenton Iron Works, New Jersey."

"And what does 't' want here!" he asks scornfully.

"I wish the kind of employment I ask for," you answer quietly.

"Then thee'll not get it here, lad," says he, rising abruptly; and, starting towards the siding in which the wagons for an empty trip are being collected, he leaves you standing alone. He turns around and addresses you again:—

"Thee arn't strong enough for a loader, or I should recommend thee to serve an apprenticeship. Good day, sonnie!" With this, he swings his huge body into one of the empty cars of the ingoing trip, and in a short time the light blazing on his head is the last you see of him, as he disappears in the drift, and is shrouded in its darkness. You feel almost relieved as you are left alone with a set purpose of getting work somewhere as a coal miner. "If the coal mining bosses are all like this one," you say to yourself, "I shall have a hard time of it." But you are not the less determined to succeed in getting work, at whatever hazard or cost. You are making up your mind to put your finer feelings into the most secure part of your nearly empty pocket book, and harden yourself for further rebuffs, if not from the boss who has left you, from some other whom you will seek at some of the other mines.

"An' is it a job you're looking for?"

You turn around to see who it is that asks the question. At the other end of a large pile of timber are two men engaged in making gangway timbers, such as you see them to be in Plate XIX.,* representing the entrance to a gangway.

---

* In Plate XIX, a set of gangway timbers is shown; two legs and a collar. Pitch pine is more particularly recommended for mining purposes than any other kind of timber. The resinous matter preserves the fibre of the wood, and prevents it from rotting. In the summer season, when the air contains a proportionately large amount of vapor, the moisture is deposited in the mine on the

"Come here and take a whiff!" calls out the same voice imperatively.

As you go towards him, the honest-looking Irishman lights his pipe, and seats himself on one of the gangway legs he has shaped out, his companion strictly following his example.

You tell him that you certainly wish employment in the mines. "Then there's all the work here that you want, me hearty. But whisper!" exclaims he, in an air of mock mystery. "There's a tavern beyant; you see it at the end of the hill. Go there and stay till the boss comes home. If you stay there all night, and have a silver bit to spend at the bar with the old woman, you'll have a job in the morning sure!"

Patrick's meaning was as kind as his words, which explained the situation at once.

various surfaces; and those substances which absorb moisture take it up to the point of saturation. When any kind of timber is used in mines which absorb moisture readily from an atmosphere surcharged with it to give it out again into an air containing a less amount of moisture than its point of saturation requires, such timber seems to be unsuited for mining purposes in all climates where the variations of external temperature are frequent, and of a considerable degree. The resinous timbers are selected, because they resist the atmospheric changes better than any other to be had in any considerable abundance.

In the sketch, the manner of setting up the timbers is shown, and also the mode of entering the hillside for water level works.

The timber is represented as being set close together. This is done in the deeper mines, whose gangways are subjected to the crushing effects of the abandoned upper levels, where the pillars have been robbed. The sets of timber are set apart a certain distance when the gangway is first opened, but in a short time the *reliefs* timbers are set up when they are as close together as they are shown in the sketch. A few laggings are shown over the collar, and behind the legs. A *main gangway*, in a lower level, will consume about a thousand feet of the best timber to each lineal yard, after it has been supplied with the *reliefs*, and as a *monkey gangway* eats up a great deal of timber also, the timber bills of coal mines form heavy items in the working expenses of them. Much of the trouble caused by the breaking of timber may be avoided by driving the main roads in the bottom rocks of the large veins, and in the top rocks of the small ones. This latter is the plan followed in countries where long wall is practised; the *gateways* are the roads cut out of the *roof*, as this is lowered to the floor after the coal is cut away.

The thickness of the timber is from twelve to eighteen inches; the laggings are from a few inches to five or six; the butts of saplings, in fact. At times there are used laggings split from logs in the manner of fence rails. When timber is set under the solid strata, care should be used in placing the wedges. These should be placed in such a manner as to prevent the weight from lying on the middle of the collar.

As an item of great expense, it is important to economize in this particular.

"The Miners' Arms" was the home of the miner boss. His wife, a thin, pale woman, working off her existence between the bar and the kitchen, was a slave to this big brute. You learn that the Yorkshireman had his own way in nearly everything at the mines which belonged to a family long residing in the city of Philadelphia. He had absolute control of the men and the work; *charged* the owners and paid the men, retaining board and beer bills, which were always of great consequence; so much so, that in a few years the boss became wealthy, and the owners of the mine poor; rich as was the mine in natural resources, and having the advantage of being bountifully supplied with stock and stores of all kinds. But what you foresaw at the hint given by the Irish timber man was realized in a few brief years; the tavern thrived, and the mine soon became bankrupt.

Of course you did not go to the Miners' Arms and take up your quarters, and you had to look farther than this for a day's work.

On the northern side of the coal basin is another large coal mine working out the coal from the same mammoth vein, having an anticlinal dip and another outcrop.

At this you find the head boss in the carpenter shop driving together the frame of a car. He drives the piece home over its tenons and critically examines the fit. The piece completes the base of the car, and he turns it over to bore holes for the bolts of the axle-boxes. Satisfied with his progress, he salutes you with a familiar nod, and encourages you by an inquiring look. You state your object.

"Yes; there's the breast No. 8 vacant, you can have that," he says; "and if you are blessed in the possession of a good looking butty," he adds, pleasantly, "you can come to work any day."

You wait until he puts the boxes of the car into their places on the frame, marks off the holes and bores them through the sills, drives the bolts up through and secures the boxes in the recesses already cut into the frame to their proper depth. The car is to be similar to that shown in the drawings, Plate XVIII., giving a front and side elevation.

This done, he springs on to the carpenter's bench and takes his seat near to where you stand.

He takes a cigar from his pocket, which you cannot accept when it is offered;

but lighting it himself, when another gentleman in a blouse, having a red head and very red nose, and being very much deformed and crippled by some accident, comes in and leans against the bench near you.

"Have you got board?" the boss asks of you, without taking the slightest notice of the red-headed man in the blouse.

You answer in the negative.

"You can take the young fellow over to your house, Dave," says the boss, addressing the man in the blouse, whose odor savored so much of that so familiarly associated with men whose occupation is to saw timber and work much in sawdust, that you conclude the new-comer to be a timber maker similar to those you have encountered at the colliery on the other side of the valley or *coal basin*.

"We have plenty of room over there; but the missus is out just now," says Dave, in answer to the boss's remark or order or interrogation, you do not know exactly which. "You can go over and see what Ann says," he adds, "and tell her you will be at supper."

Ann is the "girl," and seems to possess the confidence of the "missus." You see that Dave is an importation from South Wales. Dave pulls a dirty black pipe out of a hidden receptacle in his blouse and stuffs it full of bird's eye. And the pipe goes into full blast at once. The stem fits in a pair of grooves worn in the protruding lips by long usage, which reminds you of the grooves of a pair of guide rolls used to roll telegraph wire. As you do not go to see Ann, the boss again engages you in conversation.

"What have you been accustomed to do?" asks the boss, scrutinizing you in a critical manner.

"I have been brought up in the mines," you answer.

"Yes; but you have not worked as a miner," he says, rather pleasantly, as if to give you a chance of speaking for yourself.

"I have worked harder than miners often do," you respond, hardly knowing how to answer him.

He sees that you hesitate, and asks you point blank where you worked last.

"At the Trenton Iron Works," you answer, regardless of consequences.

"Ah! that's where our blacksmith is from. Go and bring Charlie here," says the boss, laughing.

Dave comes back with the information that "Charlie will be here in a minute."

"You can work in No. 8 nevertheless, and if you like it long enough to get used to it, I have no doubt you will do as well as the rest of us." The boss tells you this to encourage you, apparently.

You know the blacksmith as soon as you see him. His face brightens as he runs to you with an extended hand. He addresses you respectfully, wondering what has brought you about the mines. You talk of old times, and he lets it out that you have been accustomed to make mechanical drawings and maps to which he has often worked. Another man has joined you while you have been grouped together by the carpenter's bench. He is of a dark complexion, with a stern but not an unpleasant face. He wears large boots, into which his pantaloons are stuffed to keep them out of the mud, which covers those boots so thickly. He has a lamp burning on his miner's hat, and the blouse he wears covers well his underclothing and protects it from the smut and dust with which it is so thickly lined.

"This is the *inside boss*," says the head boss to you, indicating the man with the lamp on his head. "This is a young fellow I want you to take especial care of," says the head boss to the inside boss, in a manner which showed that he meant all he said against any or every objection the boss should make. "He will take No. 8 breast," he added.

The inside boss pulled at the muscles of his countenance, which was slightly distorted thereby. He did not look over-well pleased. It was with lowering eyebrows and a slightly curling lip that he allowed a remarkable English sentence to slide over his tongue, every word of which seemed to be measured by a scale, and weighed by a balance and toned by a peculiar *revolving* inflection impossible of imitation. "Understand—young—man—we—want—no—one—to—work—here—but—miners!" The head boss looked at the inside boss, and mimicking his style said, "Bill, this young fellow has No. 8 breast, and if you have a butty for him, then send them to work there!" "Very well," says Bill, "as *you* say." This in a manner which showed that he was washing his hands of some serious responsibility. Bill spoke with an accent purely Cornish. He addressed the head boss familiarly as Harry, whom he informed that the No. 8 breast was blocked up. "Well, then, be good enough to get it drawn down as soon as possible. Come to work to-morrow morning, and tell the blacksmith to get you the tools you need," he adds, turning to you. Harry's cigar is finished, and he goes back to his work on the drift-car, which he appears to be doing for the sake of keeping himself occupied.

Dave, whose pipe is glued within the grooved channels of his lips, gives you the necessary directions to find his house and Ann along with it, which is easily done. "Ann" you find feeding a young fellow about to go in on the night shift, and Ann's accent is that peculiar to North Wales, a discovery you make when she tells you that missus will be in, and supper ready in an hour and a half.

You go out to survey the woods and fine specimens of timber there are, which seem destined for the ignoble purpose of being buried up in the coal mines.

Returning, you speculate concerning the strange life you are about to enter, and wonder what kind of a woman this "missus" is to whom you have been referred so often. Is she little and ugly like Dave, and as old and crabbed as he appears to be? But you find a fine-looking woman, whose presence is commanding and motherly at the same time. There is not an ungenerous or stinted feature in her countenance. And she is looking for you as you come from the woods, to give you a welcome to her miners' boarding-house. You remark that she forms the greatest kind of a contrast to Dave, her husband.

She seems well pleased with you, and bestows on you special favor by rooming you with a young man who has charge of the steam-engines of the colliery. Thus you become fairly installed in your boarding-house.

The writer here begs to say a word for himself, and on his own responsibility.

The art of mining has been always treated in the most serious and grave manner. He begs to be allowed to treat the matter according to his own method. As for the systems in use in the different mines, and the modifications of those systems, the writer has taken much pains to obtain in the mines themselves the necessary information to deal with them. He, therefore, wishes to avoid the usual compilation of matter and maps, statistics and tables found in other publications.

The present part will be confined to the one prevailing method of getting coal at the present time out of the thick coal seams in the anthracite coal fields of America, when such coal seams are found inclining at an angle of 45° and upwards; and he will consider it his duty to give his readers faithful descriptions. If he salt those descriptions with a little criticism, and pepper them with sarcasm enough to render them wholesome, then please don't blame him. There are two grand points which he will keep steadily in view. Those are concerning the risky nature of the two kinds of investments necessary to work a mine, the risk to capital on the one hand, and the risk to life and limb on the other. Both of them suffer terribly in coal-mining speculations. In the preamble, the principal characters to be used as mediums are already introduced to the reader, and if the bosses of a mine are not the most competent to teach us concerning the manner of working it, to whom may we look for practical instruction? If we get it out of crude material, then the writer begs to say again, the matter will be only according to the original copy. The chief aim is to follow closely the work of the miner.

CHAPTER XVIII.

COAL FORMATIONS—DEPOSITS—UPHEAVALS.

A GREAT many theories concerning geological formations have been advanced by the eminent men who have spent their days in the special study of this science. Before we can describe a coal mine we must, in a limited way, examine the nature of the coal strata in which the mine is located.

The whole of the coal formations have been *deposited* while the crust of the globe in their vicinities has been, as far as violent disturbances are concerned, in a state of *comparative* or relative rest. And that this state of local rest has been continued over a great space of time in the geological history, is evident by the great thickness of these formations as they are met with in the different parts of the globe. A slow, *general* sinking of the surface has accompanied the action, while this sinking or subsiding has been checked at certain times and accelerated at others. A number of the individual beds have been formed very closely to the surface of the then existing lakes, as most of them are lacustrine in character; the coals above, the slates and rocks below the surface of the waters. The rocks and slates are deposits of the waters; the coals are *deposits of the air;* or at least they owe most of the ingredients contained in them to the vapors and gases usually found suspended in the air; and the beds have been supplied only to a certain extent with the earthy ingredients which form their *skeletons*. It is very evident that at the last the whole of the deposits have been sunken to a considerable distance under the waters; while at a considerable depth they have been broken up and elevated, and it is likely that the disturbance produced at that period (and subsequently also while the formation has been rising to the surface) has in a great measure influenced the surface by the mechanical action of the torrents created with each successive geological wave. After those disturbances have taken place, it appears that the strata have been raised up to the surface; and in the shape we have them presented to us to-day with only slight modifications. Could we have maps of those formations made from actual surveys before the measures were submerged, to compare them with the maps of to-day, a few of the localities would present very important alterations in a mechanical point of view. In some places the beds of coal have come to us in good condition lying nearly horizontally; in others the strata have been broken into fragments, if we may apply such a term to patches of five by twenty miles in breadth and width; for such we find to be in the anthracite coal fields of Schuylkill and the adjoining counties.

It is not my purpose to advance any theory concerning the past of the *coal* formations farther than that which is necessary to give my readers an idea of them in their present condition. One may well imagine the effect of a force acting on any substance possessing all the qualities of resistance that the forces of cohesion and gravitation accord to solid substances.

The weight of that portion of the globe including the coal measures can hardly be approximated except by the most careful consideration of geological principles. But, for example's sake, suppose a shell of the earth's crust five or six thousand feet in thickness extending over a considerable portion of the globe, to be acted on by a force acting in opposition to its combined forces of gravitation and cohesion. If the opposing force ever attains such proportions as to exceed the combination of forces just named, a movement will be the result. This movement will be continued as long as the active force continues to be in excess of the others. For the sake of illustration, let a bar of iron be securely fixed at each end only, so that no movement at those points can be possible. Its force of gravity will act, and there will be a certain amount of deflection, which will be greatest at the centre of the bar. There will thus be a variation between the line following the centre of the bar and the right line drawn between the central points of each end, and this difference will be due to the elasticity of the bar. In this position the strain is brought to bear on the cohesive force of the bar, and there is no more movement until by a long-continued action the cohesive force becomes less. If, however, force, other than that due to the gravitation of the bar, be brought to bear in sufficient quantity, the cohesive force can be overcome, the bar torn asunder and disposed of according to the nature and action of the forces. We have presumed the supplementary force to co-act in the same line with gravitation. If such force were to act in a line opposite to this, then before any movement could take place, that force must exceed that of gravitation, and then the movement would be limited until the force so acting exceeded that of the action of gravitation and the action of cohesion combined.

Then the bar would be torn asunder, and its parts would be elevated into a vertical or inclining position, according to the extent of the movement and force. If the moving mass was encountered by an opposing force equal to the one giving it motion, the movement would cease, and the parts of the bar would become fixed, and lie anticlinal to each other.

The coal strata, and many other formations as well, have been subjected to the same character of action, modified in endless variety; and they have been elevated at the points where the ruptures have occurred by the action of subterranean force; and being divided into sections, their edges have been tilted up to form the tops and ridges of our high mountains.

We, therefore, find the strata to be disposed in every possible position, relatively with their planes of elevation and dips. So we expect to find the beds of coal lying at all angles, from a few minutes in some localities, to ninety degrees in others.

There is one feature which takes a very prominent part in all coal fields, that is of the utmost importance to the miner. I refer a second time to the cleavage or crystallization of the measures.

Just another word in regard to this, before we go to work in the coal mines.

The main lines of the cleavage of coal, and, indeed, of the rocks of the coal formation are long; in some respects, so like the grain of timber that the faces have been (as we have seen) termed boards, and the fractures the ends of the coal. The coal splits with the greatest ease in the main lines of cleavage. It is also noteworthy that the lines of the synclinal and anticlinal axes run very nearly parallel to the main lines of cleavage, and that the principal ridges of mountains show a tendency to continue in this line. The lines of "dip" of the anticlinals consequently approximate to a right angle to those main lines; but these lines of dip are modified by the dips of the anticlinal and synclinal axes themselves. As these dip in almost every case at a rate under 15°, the changes of the *direction* of the lines of dip do not often vary more than a few degrees; therefore, as the main lines of cleavage run nearly to the east and west, the dips are mostly to the north and south. The ridges of mountains and the valleys bearing the rivers and creeks are mostly east and west, crossing from valley to valley by means of the gaps and passes which break the mountains at intervals of a few miles apart. This principle carried out, forms an extensive feature in the topographical disposition of the country. The ridges run several hundred feet above the level of the valleys, and carry the coal beds away up into the air, to use a figure of speech, and they are thus placed within easy reach of the miner whose operations, so far, have been of the most simple order. We now refer to our anthracite coal fields of Pennsylvania.

CHAPTER XIX.

MINING OF COAL—MINERS' TOOLS—STARTING THE SCHUTES—DRILLING AND BLASTING.

A MINING suit, a lamp, a set of drills and picks, constitute your outfit. With this rig you present yourself at the mouth of the drift, ready to start work. The inside boss tells you that you cannot work in No. 8 breast; you may *start* the schute, which is blocked up with big coal, requiring blasting.

Plate XIX. shows the drift's mouth near the repair shops, and near the foot of an *inside* engine plane which conveys the coal as it comes from the mine to the coal breaker, the coal being drawn up the plane by a pair of horizontal engines. We will now enter the drift, which we find to be timbered in the manner shown, with timber not less in any part than twelve inches thick; length of collar between notches seven feet; length of low side leg ten feet, and of high side leg, whose foot is set in a hole cut in the bottom rock, about eight feet. Space is provided on the low side for a water-course, as it is shown on the left of the railway, Plate XVIII. The space between the sets of timber is covered by stout laggings, as shown by those you see behind the legs and over the collars. It is often necessary to pack the spaces behind the timber by blocking and wedging, as shown in the sketches. It is a bad practice to block the collar and wedge the blocks down on to the central part between the legs, because when weight begins to exert itself on this point the collars break there with the greatest ease. Close lagging and *loose* packing are best for the central parts of both collar and legs. However, we do not find that these principles are to any great extent carried out. Miners driving gangway by contract use the readiest plan to lag and block their timbers, and we find the gangway timbers set apart from each other at a distance of 4' 6" from centre to centre, so that if it be necessary at some future time, there is room for sets of relief timber to be set between those already in.

When timbered and lagged, as shown by the plates already referred to, a gangway does not look a very unsafe passage, and it is not often that accidents

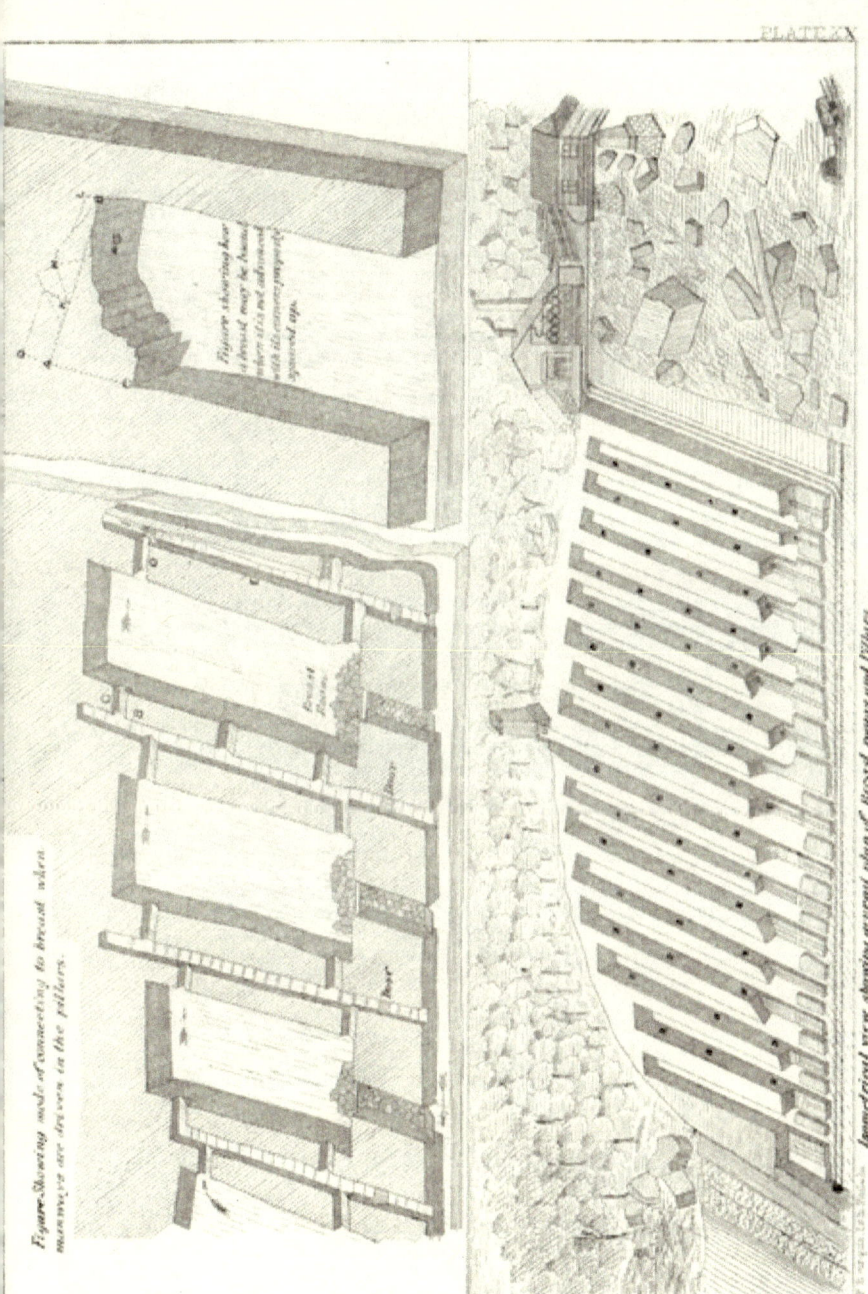

occur in them from falls of coal, or from portions falling from between the laggings. So you pass on to the foot of the slope, and then to the first opening, which is a schute leading to No. 1 breast, Plate XX.*

* Plate XX. brings us to a plan of the coal mines as they are worked in the thick coal seams of the United States. For the information of our brothers of Europe, we must state that the mines are leased by the ton and not by the acre, as mines are leased in other countries; hence a reason for the primitive modes in use. We must also state that these mines are not as yet managed by mining engineers, against whom a stubborn prejudice exists among those practical (?) men holding the executive positions. A corps of surveyors we find in the region; but these seem to have nothing further to do than to locate the mines at their commencement and survey them as they progress. Consequently the engineers busy themselves with their maps and their breakers, while the boss miner—who in nine cases out of ten (to save the credit of Americans) is an importation—ruins the properties of the land-owners by the honey-combed system of his mining operations. By the plan, we see that the breast-rooms alternate with pillars of great dimensions. The pillars are laid off to be eight yards wide at their bases, and as they are never "squared up," they are twice this width at the top. The excavations or breast-rooms are laid off to be driven where the strata are in good condition at a width of twelve yards. This is on the bottom bench. On the top, where one winged projection succeeds another until the top rock is reached, the width becomes gradually contracted until it is reduced to eight yards. Then, at this rate, under the very best of circumstances, one-half of the coal is left in pillars. Now, as the mammoth vein contains upwards of sixteen thousand tons of coal to the acre, we may see how much of it is left in the ground; because no pillar can be successfully taken out lying on a high inclination after it has been surrounded by such enormous excavations. After a property is worked out and the total quantities of the mined spaces are taken off the maps and the actual yield compared with the actual quantity of coal originally in the vein, it is found that there is seldom a yield of one-third of the entire amount. This is a severe loss to the land-owner. In many cases, owing to the large yield of 5000 tons per acre, he does not notice the loss he sustains; I say loss, because to finish the work of mining in those pitching seams, the pillars are cut through at the first headings, so as to connect the two adjacent spaces or breast rooms, where the coal and the rock all tumble down, and become so incorporated with each other that the rocks in a short time come down and choke up the batteries, and this finishes all operations. In this way the pillars are cut off from all connection with the gangway; a pretence is then made to rob back the gangway, and as long as the coal comes from the stumps above it, the miners load what they can get of it. Every now and then a crush comes on and closes up the gangway for a considerable distance in the rear of the parts operated. This crush brings the fallen rock to the front after a certain amounts of the coal which has been crushed out of the pillars has been loaded and sent to the surface, when the coal left inside of such rock is abandoned to the chances of the future. On account of the inclined position of the measures and the size of the excavations, the spaces can only fill up by an incorporated mixture of the rocks and the slates and the coals. The bottom rocks, however, will in most cases remain entire, and it is to be hoped that some of the coal in those partially worked seams may be reached by cutting "gateways" through them. The figures accompanying this sketch give the main features connected with the airing and working of the places in detail. The figure to the left at the upper corner is a plan on a flat surface, showing the mode of progressing with a sheath of breasts. The arrows point out the direction of the ventilation through the headings and manways.

A schute is a passage turned from the gangway at right angles, and it is driven about nine feet wide and five feet high and timbered, as shown in Plate XXI.* The instance shown is that in use sometimes where a connection is made between the room of a breast and the gangway by one schute only, instead of by two. The pitch of the bottom state is at an angle of 45°. You will observe that the schute starts from a spacious platform a little above the level of the gangway wagon. The

The position of a *monkey gangway* or counter air-way is shown in section in Plate XXIII. This monkey gangway serves the purposes of a return air course, until it is shut up by the crush. After this takes place, the air is left to get out as best it can by way of the breasts, and by the upper levels. When coal below the water level gangway is being worked, it is reached by means of a slope sunk down the incline; and to an air-passage driven alongside of the slope, the monkey gangways from the east and west portion of a property connect. To the gangway, cross holes connect the monkey, which may be driven over the main gangway, as it is shown in Plate XXIII., as well as alongside of it. In a manner the monkey, together with the main gangway, answers the same purpose as do a pair of *headways* in an English mine. The variation of this method, when used to take out coal in veins slightly inclined, consists only in changing the direction of the breasts, which are occasionally driven in a manner diagonal to the direction of the level gangways. This system of driving excavations in England is termed *cross-cut*. It is not to be recommended anywhere for any purpose whatever, because planes can be easily arranged to run coal from the breasts, if driven at right angles, and light wagons may be used for this purpose. But this comes under the head of transportation, which is a subject of secondary importance when compared with the primitive one of mining; so we cannot compare nor criticize the various modes of arranging the breasts to suit and adapt them to different modes of transportation. Matters like these are left to the sagacity of the men in charge of the mines, and they vary them to suit their own ideas.

* Plate XXI. gives us the manner of connecting together the main gangway and breast-rooms, and shows us the manway by which the miner climbs up to his work at the breast. The loose coal is represented as blocking up the starter's battery. To the right, the top of the bottom bench is shown, with a portion of it removed, to show more clearly the mode of placing the timber, and of planking up the schute. The steps over which the starters climb to the battery, with a portion of coal lying on the platform, are also shown. The place for the starter is at A. On the left, we see the gangway under a section of the upper benches of coal. The upper portion shows the coal pile lying above the battery, within the breast-room. Schute timbers are about nine inches to a foot through, excepting the battery collar and centre legs, which in some cases of heavy inclination are often eighteen inches. The steps in the manway are set into the sides by cutting holes, into which they are wedged at a distance of three to four feet apart. The cross-headings are driven to the right and left into the breast-rooms on each side of the pillar, and the distance apart of the headings varies from twenty-four to thirty-six feet. A manway door is placed at the position A, and it is used to intercept the air-current, and force it into the most advanced works. This door is shown on a large scale by a figure in Plate XXIII., and it is made to open outwards, when coals running down the manway strike it sufficiently hard to knock it open; it closes automatically. A platform is built at the bottom of the manway, which collects the coal cut in the face of it, and in the headings as these are driven over into the breast-rooms.

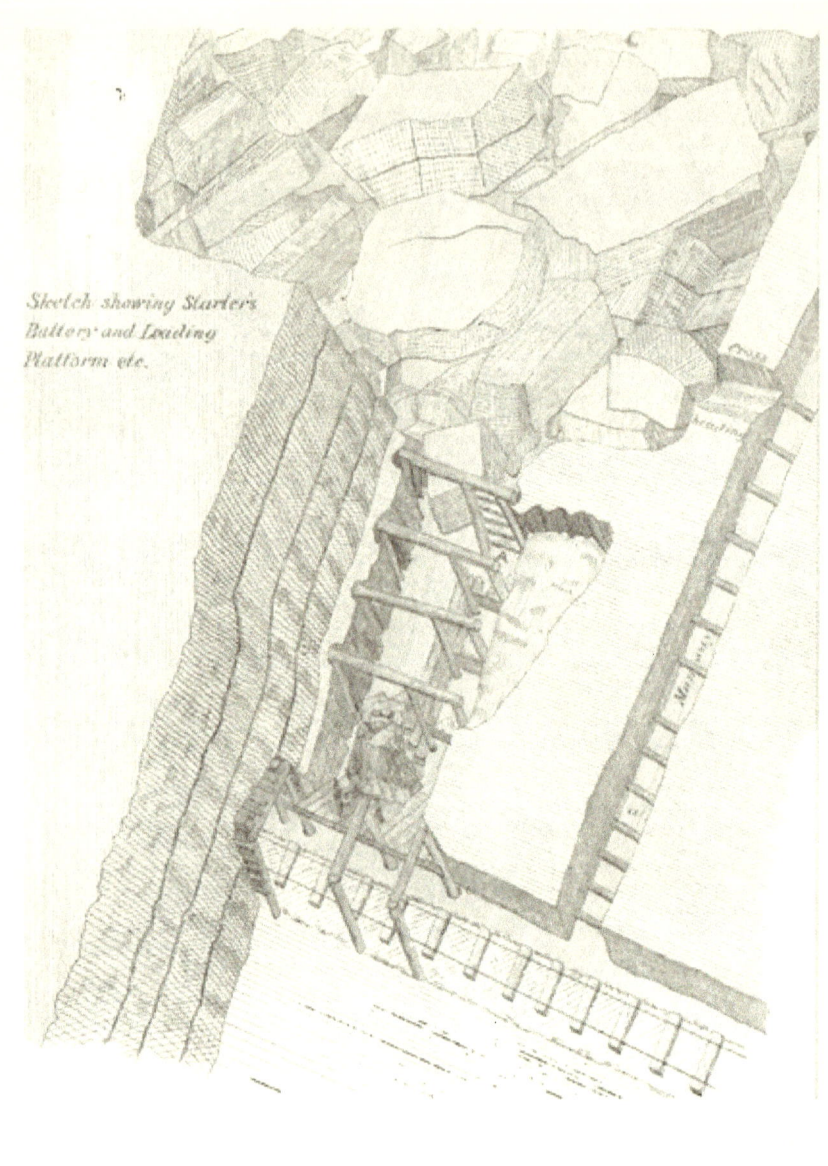
Sketch showing Starter's Battery and Loading Platform etc.

PLATE XXII

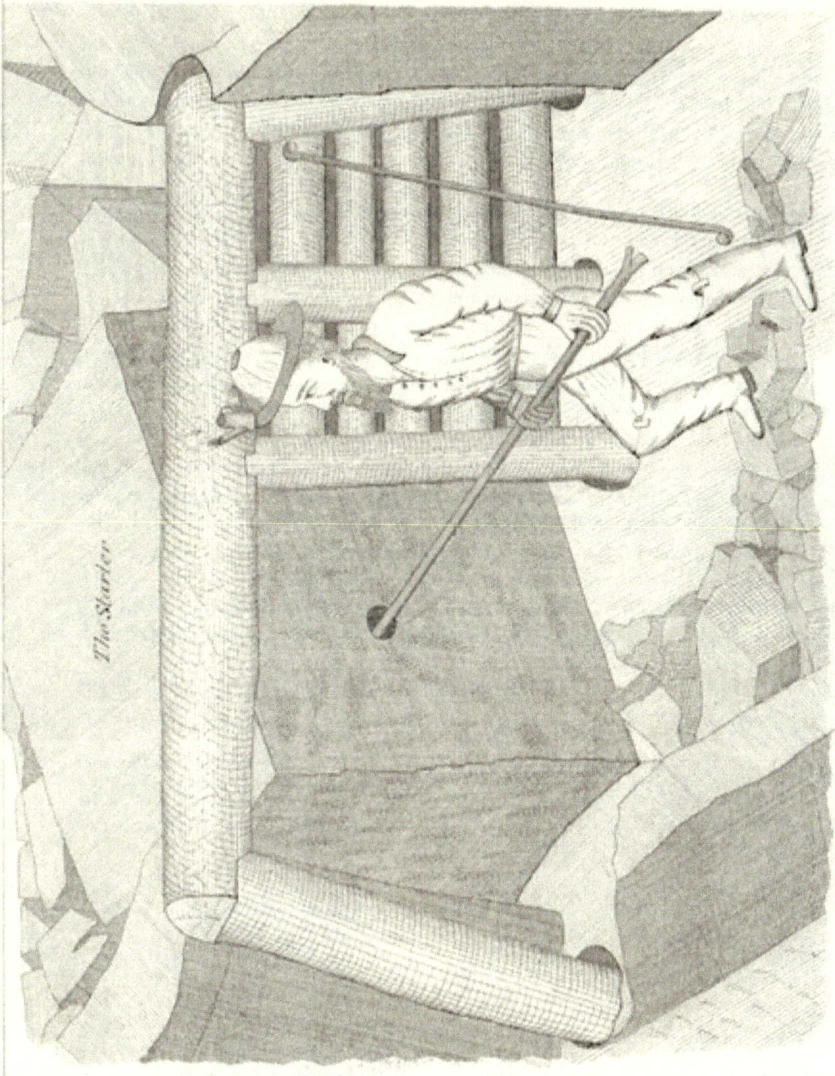

The Starter

collars are furnished with centre legs, to which the schute planks are nailed. This partitions the schute off into two unequal parts, the largest one being reserved for the coal that comes through the battery shown in the upper part of the schute at A. The upper set of timber is very stout, the pieces forming it not being less than fifteen inches thick, and it is set into the coal at each end, so that there can be no possibility of its slipping out of place, which would be a serious mishap. The smaller space is provided with steps, and reserved for a travelling road, so that a "starter" may go to the battery, and break the large blocks of coal, as they wedge in between the battery timbers. This battery collar is generally provided with two centre legs bolted together to give them greater power of resistance, and enable them better to oppose the heavy shocks and severe strains to which they are subjected; and any one can form a tolerable idea what the strains on the battery collars will amount to when he knows that a breast is driven up the incline to a distance of three hundred feet, of a width of ten to twelve yards, a height equal to the thickness of the seam, which is sometimes thirty feet. The battery, as it is now shown, is *blocked*, and is started by a man standing near A to bar or to blast the blocks of coal wedged between the legs. Plate XXII.* shows the starter at work. When the coal is "started," it runs down the schute on to the platform, and if the blocks of coal in the breast-room are less than the space between the timbers, they crush through the battery, and keep the schute full, while the loader rolls the coal into the wagons at the platform by barring the large pieces and shovelling the lesser ones. It is not wise to empty the schute entirely after the battery becomes blocked, because much of the coal, when started, would then run over the edge of the platform on to the track, and cause unnecessary delay. By this time you have an idea what a schute is, how it is timbered, and how it is started; so all we have to do with the schute at present is to recount—for an example—your personal experiences at schute starting.

A battery is to a schute what a regulating valve is to a supply pipe. The breast-room or excavation is as it were the reservoir.

* Plate XXII. gives the battery collar with the support of an extra leg under the battery. It shows also the starter at his work, and how the ends of the collar and the side legs are set into recesses cut into the coal to prevent their being driven out of their positions by the crushing effects of the coal passing down the schute. The starter's duty is obvious; he breaks the coal at the battery to get it into the schute below.

The schute of No. 8 breast which you are ordered to "start" by your friend, the inside boss, is what the miners term "blocked up" at the battery, in the manner shown in Plates XXI. and XXII. The schute below the battery is partially empty. The travelling road to the right, shown in Plate XXI., which leads from the platform on the right of the gangway to the battery at A, gives you access to your work, as you climb with your tools from one step to another until you reach it. The battery is securely blocked, sure enough, and there is no danger of anything "starting" until you drill a hole in the lump of coal which forms the key of the blockade; and not then, until you fire a blast in the hole you have drilled. But blockades in batteries do not always form so favorably. In some cases, the mere touch of a crowbar will "start" the coal, when it rushes down into the schute with a crash, carrying away the bar out of the starter's hand; and unfortunately, sometimes, a starter goes down along with the coal. It is an extremely dangerous operation to "start" the coal into a schute, when the blockade forms a few feet above the battery by there arching itself. Then, a piece of coal, or the crowbar thrown so as to strike some particular piece composing the blockade, may move the whole mass of loose coal in the breast-room, part of which runs into the schute, the balance being retained by the battery, which often receives its weight with a severe shock. While this is taking place with the loose coal rattling about his ears, the starter depends on the stability of the battery for his life.

Before we detail your mode of drilling the hole in the block of coal, already referred to, we shall describe a set of tools used to mine coal in the breasts of those highly inclined coal beds in Pennsylvania.

Fig. 1 is the drill, as it is commonly made. It is usually made about six feet in length, and has one cutting edge, which here is shown to be a straight chisel two and a half inches long; but some sharpeners of drills file its edge in such a manner as to allow the corners to project ahead of the central part. The chisel is then compounded of two straight parts, forming in the middle where they join an obtuse angle.

The other end of the drill is set up to form a butt. In this a groove is made to slide along the "needle" (Fig. 2), while it is used to ram back the tamping.

Fig. 2 shows the extremities of the needle. The needle is made in some cases of good tough iron, three-fourths of an inch at the eye, and it tapers evenly down to a point, at a length of four to six feet.

Fig. 1.

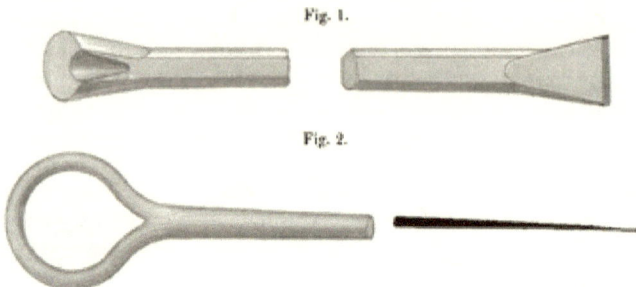

Fig. 2.

To clean out the hole, and prevent it from becoming clogged during the process of drilling, a scraper (Fig. 3) is used. This is made of one half-inch round iron, and it is generally about six feet in length.

Fig. 3.

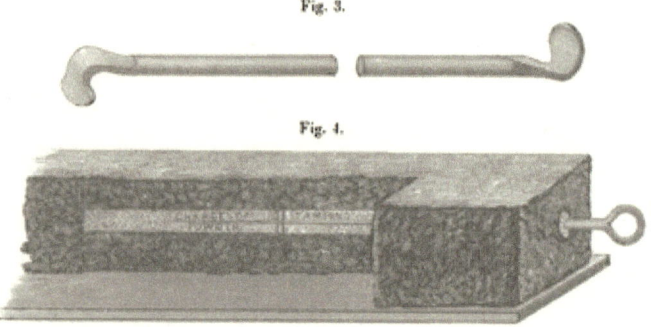

Fig. 4.

When a hole is drilled in coal to any required depth, after it is carefully cleaned out, the charge of powder, made up into a cartridge, with the point of the needle sticking in it, is inserted in the manner shown in Fig. 4. Then the tamping is rammed tightly into the hole to fill it up to its mouth. After the needle is carefully drawn, a tapering hole, extending from the mouth of the drill hole to the charge of powder, is left through the tamping. The hole is then ready to receive the squib and be fired.

The squib (Fig. 5) is made of a piece of straw or a cylinder of paper filled with powder, one-sixteenth of an inch in diameter, and about four inches long. One end

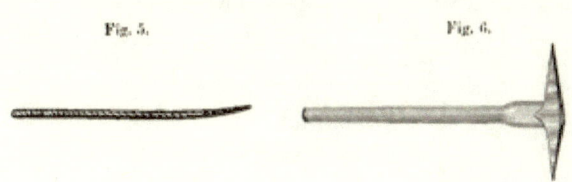

Fig. 5. Fig. 6.

of the squib is made to form the match, which is made to burn sufficiently slow to allow the one who lights it to retreat to a place of safety. Before inserting the squib, the sealed end is nipped off. It is secured as loosely in the hole as possible, so as not to prevent it from flying through the hole, on the principle of a rocket, to ignite the charge.

When a drill hole cuts a feeder of water, a piece of gas-pipe one-fourth to three-eighths of an inch in diameter is used, instead of a needle. In this case, the cartridge is made water-proof; and the pipe, termed a "blasting barrel," is tied into one end of it, in such a manner as to form a water-tight joint. It is then placed in the hole, and the tamping is rammed back in the same manner as if the needle were used. A piece of small wire, kept within the "blasting barrel," until the tamping process is completed, is then drawn out. This disposes the grains of powder in a favorable position to be fired when the rocket squib flies through the barrel. Without the use of this wire, the grains of powder become jammed in the barrel in such a manner as to prevent the fire of the squib from igniting promptly the powder of the charge, which sometimes results in serious mischief by causing the charge to hang fire, and to explode at the moment the miner returns to examine into the cause of the delay.

MINING OF COAL.

In Fig. 6 is shown the style of pick mostly in use in the anthracite coal region. They are made of weights to vary between two and seven pounds, the lighter picks being used to dress the sides of the manways, to undermine, etc., while the heavy picks are used to raise and pry up the large blocks of coal loosened by blasts, and to break them into pieces of a size convenient to be handled.

With a steel sledge of six to eight pounds in weight, and a steel wedge about six to eight inches in length, made to correspond to Figures 7 and 8, the miner has a complete set of coal-cutting tools for the hard anthracites.

Fig. 7.   Fig. 8.

Returning to the battery, shown in Plates XXI. and XXII., you drill a hole in the lump of coal blocking it up. You use the drill described in Fig. 1.

The usual method adopted in drilling is to fix on a point to commence the hole, and decide on the line of its direction, aiming at another point as its termination. Let the object in view be to place the charge of powder as near the centre of the lump as possible, so that the force of the explosion will radiate from this point, and break it into several pieces. Therefore, you survey the lump of coal, using your eye to measure its angles and dimensions, and make your decision from the results. Your success will depend very much on the accuracy of the survey; so you are careful in these simple preliminaries. The point is selected and the hole is marked out with the point of a pick. Then you stand on the step of the travelling road, and balancing the drill, take aim at the hole, and strike a hard blow. You turn the drill a little, and aim at the same point. Both blows take effect in different places. By continuing the operations of striking the hole and turning the drill, you strike the chisel on a series of fresh points. You continue this until the hole is advanced to a depth of six inches; and if your blows have been given accurately, the hole is tolerably round and straight. Then you take the butt end of the drill in your hand, and allowing the chisel end to slide in the hole, you change your position to suit your balance, and work the drill by short, quick strokes, turning the drill a little at each blow, until the hole is so far advanced as to allow a longer stroke to be given with a correspondingly greater effect. With practice, you find that the great secret in drilling consists in being able to turn the drill in such a manner as to keep "corners" and "flats" from forming in the hole, and to acquire the skill of punching them off when they do happen to form, in spite of your care. If you do not pay attention to these very small details, you expend much of your strength and your time to no purpose whenever you strike *ahead* of a back corner. Assuming that you have the hole drilled into the point aimed for, you next take into consideration the succeeding operation, which is that of charging your hole in the manner shown by Fig. 4.

Assuming that your hole has been carefully tamped, and the squib properly inserted at the mouth of the needle hole, you put your tools in a place of safety, touch the match of the squib, and retire yourself to such a place—the gangway, for instance—and bawl out "fire!" The result is an explosion; then a rush of coal down the schute, some portion of it in this particular instance tumbling over the platform on to the railroad track in spite of the planks laid temporarily behind the legs of the gangway to prevent it. However, the schute is started, and the loader comes along with his wagon the succeeding trip.

Now a word of the loader and the driver and, the boss loader while the smoke of the blast is clearing away, and the coal in the breast and schute is settling.

At intervals, in driving a gangway, the width is increased sufficiently to double the track and make sidings. The length of these sidings is from one to two hundred feet. This allows standing space for a trip of cars. To work and load the wagons, a gang of *loaders*, headed by a loader boss, is employed. If the distance is too great for the loaders to work the wagons from the schutes and platforms to the coupling-up siding, a bumping mule is furnished. This bumping mule is trained to bring out the innermost wagon and join it up against the next in advance, because, on account of the schutes being arranged at equal distances from each other, the wagons are kept at least this distance apart from each other while they are being loaded. At times there is barely room for the animal to pass the cars in advance of him, and it is often serious, and sometimes fatal, for both driver and mule should the gears become entangled among the timbers or the framework of the wagon while the mule is in the act of siding past on the gutter side of the gangway.

A certain number of wagons drawn into the siding constitute the trip to be taken out to the foot of the hoisting slope, and in cases of water level drifts being above the dumping schute of the breaker, to the surface. The number of cars to be coupled together depends on the condition of the road, and the amount of its grade. You often see a team of four mules bring out a train of cars bearing forty tons of coal.

When the loaded wagons are bumped and drawn into the siding, the empty ones are distributed among the breasts which have coal ready, and their schutes are "started," or full of broken coal. A glance at Plate XXII. shows that the platform on which the loader stands is higher than a drift wagon; therefore the coal is very easily shovelled into it. When the inclination of a coal seam is lower than 20°, the platform is not so high, and the coal must be broken into pieces sufficiently small to be handled and thrown into the wagons. This is hard work on the men's hands, and the sharp edges of the coal cut terrible gashes, if they happen to be handled carelessly. While being broken, splinters of coal fly and cut any unprotected portion of the body they strike. A man may only be aware of his wounds when the blood moistens the handle of the pick or instrument he is working with, and renders them disagreeably slippery. So much for the loader when he is called—or rather sent by the boss loader who is a kind of "despatcher"—to the schute of No. 8 breast.

Of course, the road must be cleared to begin with, and on account of this little extra labor our loader has his little "growl." The rest of the loaders have their laugh and their joke.

"Hurry up, now, Mickey, and get that stuff out of the way," says one of the crowd, waiting.

"You can go to thunder, Pat," retorts Mickey.

"Come on, now," adds the boss loader, lending Mickey a hand to expedite the work, "let 's get the thrip in as schoon as we can."

"Get the thrip in yerself," says Mickey, working like a good fellow to stop the gabbling and jeering of his comrades.

The road is cleared sufficiently for the wagons to pass in by putting some of the coal into the first empty wagon; but more is left in the middle of the track, and on the gutter side of the gangway, where there is more room. There it remains, to cut the feet of the bumping mule until it is better cleaned during the night. The trip passes in and the wagons go to their respective schutes, and are duly loaded, a little competition taking place during the operation, and a bandy of words at the end of it.

"Is not that wagon loaded yet?" asks the first man finished, of the next man to him, putting the last lumps of coal on the top of his wagon.

"Be me soul, now, it's not loaded you are yerself, so stop yer everlasting jawing," is the retort. And so on, during the day, the dreary work in the mine is enlivened by these and kindred exchanges of compliments where the best of feeling in most cases prevails.

The first day is passed over, and to keep No. 8 schute started required the aid of seven blasts, the length of the holes varying from fifteen to thirty-six inches.

Although your friend, the inside boss, has not interfered with you, he is satisfied with your day's work; and as the breast-room is still blocked at the face, you are required to continue at the *starting* the succeeding day. But the third day you are allowed to proceed with your work in the breast. In the mean time a "butty" has been found, with whom you are to form a partnership at the work of No. 8 breast.

CHAPTER XX.

DRIVING A BREAST—COST OF COAL—MANWAY AND HEADING—BLOWING DOWN TOP COAL.

As you will now proceed to the face of the breast and begin to work the coal there in the manner you know best, your object will be to advance the breast as rapidly as possible up the incline of the coal seam.

The manway through the innermost pillar forms the travelling road for No. 8 breast. You mount the platform, and about nine feet from it, come to an air door (shown in detail in Plate XXIII.*), which is hinged on the top of the frame and made

---

* Plate XXIII. gives us a view of a mode of timbering the manways in the breast-rooms, and of connecting them together by means of headings driven clear through the pillars. The manways connecting to the breasts start off through the stump over the gangway, and they are headed over to the right and left to connect with the breast-rooms on each side. At the bottom of the breast-room batteries are formed by placing one of the ends of two stout stringers on the battery collar, as shown at A; while the other ends are put into recesses cut in the bottom slate. Laggings are then stretched from one to the other. This forms a space underneath, in which the starter finds room to start the battery amidst the cloud of dust raised by the coal as it crushes and grinds through the battery. The road into such a battery is the one already referred to. The manway within the breast-room commences at the battery, and is built of props six to eight inches thick and of planks six feet long, in the manner shown. The props are called jugglars, and are about five feet in length, and are set about as many feet apart. The planking laid and spiked on them from the bottom rock up to the point at which they lean against the rib is formed of two-inch plank. As the breast advances up the incline, the rooms fill up and cover those manways; and thus the miner is provided with a space through which he crawls to his work, and by which he retreats to a place of safety in times of threatening danger. The breast-room to the left is shown to be full of coal, exactly in the manner that breasts are worked by the miner. The bottom bench of the breast is shown to be advanced, and the top coal is over-hanging, ready to fall, or to be blasted down. A section is shown to the left, which shows the place of the monkey gangway, as it is sometimes driven over the top of the main gangway, instead of at the bottom of it. At A, the place of the starter is shown. The breast-room to the right shows the inside manway, and the pillar dividing the breast-rooms is shown to be pierced by a heading driven near to the breast of each excavation. The pillar on the lower part of the excavation shows how it may be squared, to avoid leaving the projections shown in that part of the breast advanced further than the bottom bench, up to which the manways are built in working order. The schutes at the gangway are provided with platforms, which allow the coal to run into the wagons of the gangway. From this sketch, an idea can be formed concerning the quantity of coal left after the mines have been worked over the first time.

15

to open outward to allow the coal cut in the face of the manway to open the door
by its impulse, after running down the manway. The door is made to close automatically by its gravity. In the manway you find steps about three feet apart.
These are of tough saplings, about two to four inches through, and well secured in
holes dug in the sides. You experience no trouble in climbing the manway, whose
height and width are two by three feet. "Cross headings" start off at right angles
about thirty feet from the gangway, and are driven, one to the right into No. 8
breast, and another to the left into No. 9, in the manner shown more in detail in
Plate XX. You crawl through the outer heading, and find it blocked up with coal;
then pass on up the manway to the upper heading, which is open. Through this
you enter into the breast. The heading is at least twenty feet from the face, and it
is driven in the upper portion of the bottom bench at a distance of not less than
four feet above the bottom rock. This allows an open passage to the breast, when
there is a considerable amount of coal in the breast-room. But within the "breast"
fairly, you have a scene—a coal section—presented to you, a fair section of the
Mammoth Vein as it is usually seen in the Mahanoy and Shenandoah valleys. The
mass of loose coal lying in the breast-room over which you crawl is very considerable; but the section represented by the breast is a magnificent one. Beginning at
the bottom bench, you find five feet six inches of coal glistening like jet, having
a vitreous fracture, whose shining surfaces almost reflect your visage in a dozen
different places at once. Over this you have a few inches of black slate, then a foot
of hard coal, showing a genuine cleavage. Next a band of slate, which varies from
four inches to one foot in thickness. Then come forty-eight inches of coal, containing carbon to the amount of ninety-four per cent. Above this are three inches
of slate, and then follow forty-five inches of coal, containing a couple of streaks of
bone coal, from which fact this coal has derived the name of the "bony bench." A
slate of two to four inches separates this bench from the one above it, which is
thirty-four inches, and contains a liberal portion of the sulphurets of iron. The
slate which follows varies from a few inches at the point where you have taken the
above section, to several feet in others, is termed the partition slate, and it separates
the last-named bench of coal from the top bench of forty-two inches, which latter
is not of so solid a nature as the rest of the coal seam. This is the magnificent

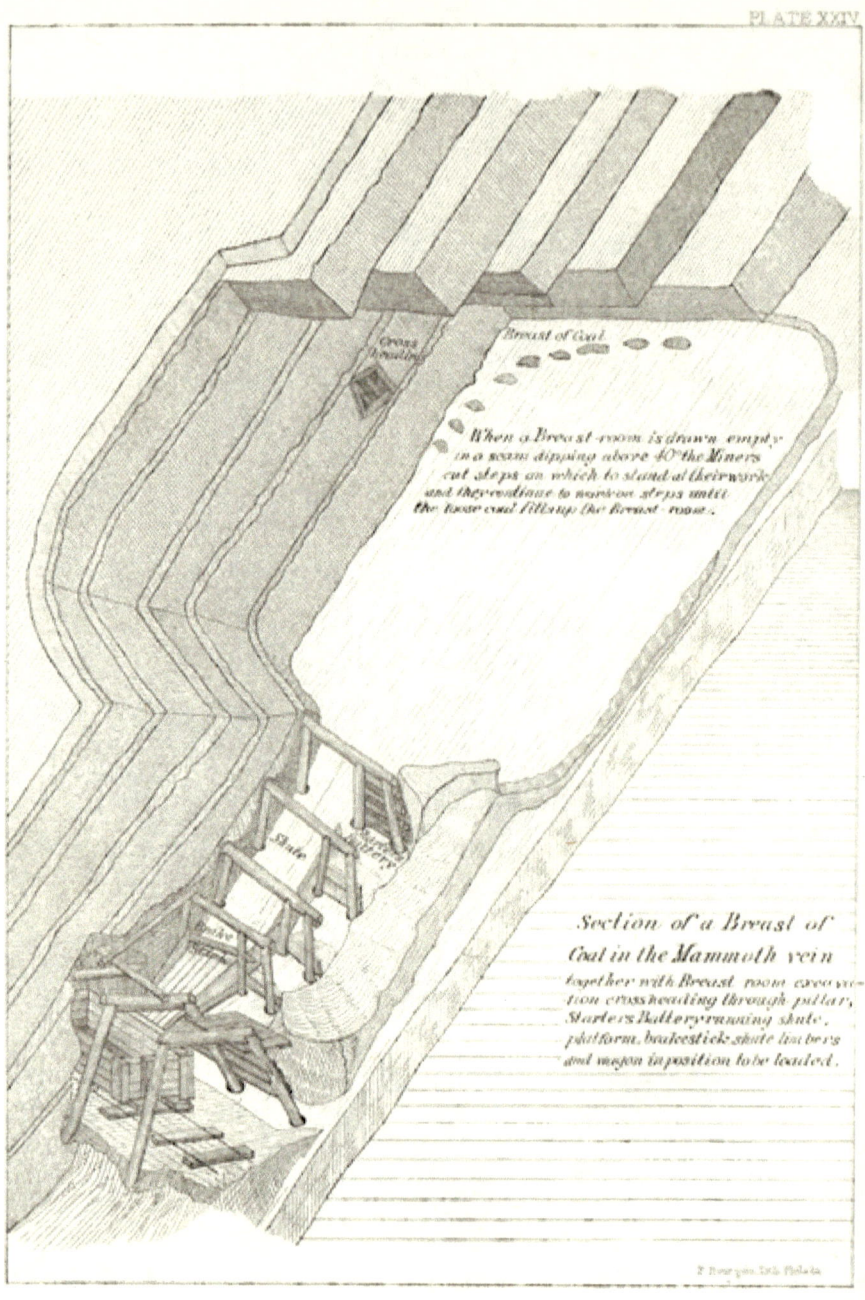

Mammoth Vein, near Ashland, and it is covered with slate, a few feet in thickness in some localities, while at others the hard rock forms the roof of the coal. The same alternation of clay slate and rock occurs as often under the seam as the slate and rock do above it. (See Plate XXIV.*)

You look at the breast and find it to be in width thirty-three feet. Such a seam of coal should yield about seventy tons to the lineal yard. Seventy tons for nine dollars, the price per yard of breast—the digging of this amount of matter only costs thirteen cents per ton! A loader loads off a high platform from forty to sixty tons of coal per day, and this adds less than four cents more to the cost when it is in the drift car. It is the dead work, or preparatory work, the timber, railroad iron, repairing, and outside work, which form the *great items of cost in the coal* fields of Schuylkill; yet the complaint of miners' high wages is often made a plea.

While you have been drawing the section given above, and estimating the costs as far as they are stated, your butty has commenced to drill a hole; and by doing so has taken possession of the *soft side of the breast*. It is necessary to explain here how a breast has a soft or free side, and a hard or tight one, and a sketch to show the disposition of the cleavage or "grain" of the coal will assist us to do so.

To the right of Plate XX. is a figure showing a breast where the line of cleavage does not intersect the line of dip at right angles; but instead at an angle of about 70°. Therefore, if the breast were driven up to the face of the cleavage, as shown at $A B$, the angle that would then be formed by the pillar of the right side of the breast with the line $A B$ would be an obtuse angle of 110°, while the

---

* Plate XXIV. shows a breast out of which all the coal has been drawn. It presents us with a tolerable section of the famous Mammoth Vein. A drift car is under the schute. A brake stick is shown, which is used by the loader to govern the flow of the coal into the wagon. A heading is also shown to be driven into the breast-room, out of an adjoining manway. It is not often a breast-room is drawn empty; but when this is done, steps must be cut up to the breast on which the miner stands to drill his holes. When such breast-rooms fill up, the miner stands on the loose coal, and only allows a certain amount of coal to be drawn daily, which will correspond to about one-third of the amount mined. This allows him a good footing on which to stand and perform his work in a comfortable manner.

The corner of the pillar at the battery is shown to lean down the hill, which admits of a view of the rib of the breast-room, which is, in this instance, shown as being cut squarely up from the bottom slate. The upper layer is the slate roof of the coal vein. The floor is shown partly in section on the right of the plate. The section in the breast is shown in a succession of steps and projections.

line along the other pillar, united to the same line $A B$, would form the acute angle of $70°$. To keep the breast advancing as nearly as possible in this line, so that on squaring up the corners with each cut taken off, the breast will be parallel with the cleavage, is certainly the most economical way of working off the cuts; and the saving is effected, both in powder and time, both in themselves representatives of ready money to the working miner. If, instead, the line along the breast be from C to B, the piece of coal forming the triangle $C A B$ is in a very tight place, and a miner working his place, thus taking his cuts off so as to keep his breast advancing parallel with the line $C B$, will tell you that he has a tight breast.

Now, supposing that two men working together at a breast agreed to cut out the coal delineated by the triangle $C A B$, to bring up the breast parallel with the cleavage, then would the breast be ready for the reception of a drill hole from G to H, the result of a blast fired in such hole would approximate the dotted line $B H K$.

If the coal were as strong as the most compact of the anthracites, and the hole drilled in the position indicated were about three inches in diameter and nine feet in length, a charge of well-packed powder, equal to four feet in length, would almost sweep out a cut of a couple of yards. The rib at B would be so much shattered as only to require the pick and wedge of the *skilful* miner to dress it up to the cleavage face at L.

On the other side, which is the hardest, a couple of well-directed blasts would dress up the corner from A to D.

To work breasts systematically thus, two sizes of drills are necessary; the larger size to be used in the middle of the breast, the smaller in the ribs. This principle of working coal is adhered to in mines worked in the bituminous coals; and, were it not, the coal would become so "woody" as to retard the work of the miner fifty per cent. Illustrate it in the following manner: Take a piece of white pine weatherboard having a straight grain. Draw two lines parallel to each other, from one side to the other. Let those lines be at an angle of $70°$ to the grain of the wood. Now take your knife and cut out the material between those lines. You will find that side forming the obtuse angle with the grain to work much easier than the other. When the two corners are cut so as to be advanced to the same line in the grain, the middle will be the more readily split off. The same practice may be carried out in cutting the coal off a breast. If the coal is not worked according to the principles shown, you will see no grain; it will show you its vitreous fracture instead, and the powder you will require will blow it into too many splinters to form the best coal. You will tell us, too, that the coal of your breast is hard, and possesses no grain, which is in most cases sheer nonsense, because coal, except it be twisted and distorted as it is near faults, always has a well-defined cleavage.

But you work on your breast in the best way you can, advancing well in the bottom bench to undermine the top coal. In the plates referred to so far, we have shown the operations to be progressing in the *bottom bench*. The Plate XXIII. shows a breast to the left with the bottom bench removed for a considerable distance in advance. The balance of the coal is said to be undermined, and it is termed "top coal." Therefore, the next operation is to blast down this remaining portion of the seam, which is by far the most important in regard to its great production of coal.

The blasts in most cases are prepared by drilling holes from the undermost sides of the benches; but instead of being drilled at right angles to the line of inclination of the seam, they are directed a little more to the dip. Of course, we are speaking of veins having a considerable amount of dip.

In veins having a low rate of dip, holes may be drilled from the face of the benches nearly parallel to the dip of the coal vein. In any case, miners contrive to reach the holes they drill in the top by allowing the loose coal to heap up under the points in the coal to be drilled. Otherwise, they would build scaffolds or stand on the steps of ladders to drill their holes, which is both dangerous and inconvenient.

In the event of the coal being very hard, it is well to blast off or pick off all the projections of the undermost benches. In some cases the top coal does not require any blasting; it falls freely as the bottom bench is removed.

When the top coal is blasted down and the breasts well blocked up and squared up, you commence to work forward the manway through the inside pillar, and finish by making the heading connections near the breast of coal. The current of air circulates among the breasts, as shown by the arrows in a figure of Plate XX. The new headways are driven from the breast toward the manway, if it is possible to enter the breast-rooms by any of the back headings. Otherwise, admission may be had to the breast by cutting the outside heading D through to the breast from the manway E. But usually by a little work, a hole may be made through the loose coal blocking a lower heading. In the mean time the loaders are busy drawing the coal out of the breast, and while you are working at this dead work for which there is no pay, you hear the roar of the coal as it tumbles down to the battery. Sometimes there is the report of a blast at the "starter's" battery, which is followed by the thundering of an avalanche. But the coal is never still while the loader is at work loading at the rate of sixty to one hundred tons per day, until the coal in the breast-room has moved so far as to allow of the resumption of work at the breast. This may require the work of a week of steady loading, and in the mean time you have your manway and heading connections made close up to the breast.

The headings are the miner's store-rooms, and each new heading finished near the face receives his stock of tools and powder, his oil, and his dinner-cans. When he sets fire to the match that explodes his blast, he finds a safe place of retreat in the nearest heading. When the top coal threatens to fall, it is his harbor of refuge.

A manway may be three feet wide by two feet in height, and it is a matter of choice whether the miner carries on the face of his manway or heading by blasting in the "solid," or whether he assists his blast by mining or "holing," a process resorted to generally in all narrow work by all good miners. Usually a thin piece of soft slaty matter lies on the top of the bottom bench, which is taken advantage of by the pickman, and with the aid of his light drill and pick, a distance of three feet may be holed, by first picking out the soft layer, and then by blocking or chopping down a portion of the coal in front, to allow the play of the arms when the distance of the holing becomes too great to work the pick freely. Miners accustomed to a skilful use of the light "mining pick" have a great advantage over those who depend altogether on the operations of their blasts in such narrow work.

After a "cut" has been holed, a drill-hole is made near the top of the manway; and after this has been properly charged with powder and exploded, the effect will be or should be to square up the face to the point to which the holing has been carried. As the inclination is such as to allow the coal to fall freely down the manway, nothing remains to be done but to start off the loose pieces hanging on the face, and dress up the corners ready for a succeeding cut. In case of blasting in the solid, the coal is only shattered, and in most cases more pick work is required to chop off the coal than would have undermined a good cut twice over.

For the ventilation of the manway while the miner is working in its face, the rushing of the coal down the steep incline is depended on for a renewal of air to displace the foul air exuding from the coal and vitiated by the burning of the miner's lamp; because, until a connection is made by the heading and manway, no constant current of air can reach the miner. It is therefore a matter of consequence to himself that this passage be completed as soon as possible after it has been commenced.

The distance between the headings may be anything between eight and twelve yards.

As we have detailed your mode of advancing with No. 8 breast, whose manways are driven longitudinally through the centre of the pillars, you will please return to the surface and come in with the current of air which finds its way inside to ventilate the works.

CHAPTER XXI.

VENTILATION—DRAINAGE OF WATER LEVEL—GANGWAY IN BOTTOM ROCK.

The ventilation of a mine worked under the conditions, and according to the method shown, is on a very simple scale indeed. The force of the current is obtained from the difference between the level of the ventilating funnel built on the top of the air-way, and the level of the mouth of the main drift and water level, which in this case is over two hundred feet. It is only a repetition to tell you on what principles natural currents of air depend for their motion; but we shall repeat the old tale.

To begin, we shall state that the pressure of the atmosphere at the top of the air-way is a certain thing at a stated time. Then the pressure of the atmosphere at the drift's mouth is this certain thing, plus the weight of a column of air two hundred feet in height. This has reference to the weight of the external air. Now, as equal volumes of air having equal temperatures at equal heights possess like weights, we have the air in the mine, and at the surface, exactly of the same weight, at the same levels at all times, when the temperatures are equal, excepting (if we are to be very nice about it) when an excessive quantity of vapor or gas exists in either of the atmospheres. In the relative conditions we have premised, the air of the mine and the air of the surface would balance each other, and remain stationary, and there would be no regular ventilating current, and no currents would move except those driven backwards and forwards by the action of the wagons passing along the gangway. This miners call baffling. Now, we know that when there is a difference of temperature between the air outside and the air inside of a mine, and the connections between the two points (that is, the top of the air-way and the mouth of the drift) are open, there will be a preponderance of weight inside or outside as the case may be, which will put in motion the body of air occupying the space inside between the points specified. As the air inside remains very nearly the same in temperature, winter and summer, the changing of the temperature of the seasons governs the current of air in the mines in direction and in intensity; therefore, in winter, the air is condensed, and possesses greater weight outside of the mine by the $\frac{1}{436}$th part to each degree in difference of temperature. Now, the increase in weight of two hundred feet of air having a difference of 50° in temperature is very easily found out, when we know that thirteen cubic feet of air ordinarily will weigh about a pound *avoirdupois* by applying rules well known to us all who have thoroughly studied our lessons of addition and subtraction. If we are not mistaken, with such a difference in temperature, we would have a difference in pressure of about one and a half pounds per square foot. This is sufficient to put a current of air in motion at a lively rate.

Now, as we have no text-books which treat this problem accurately, I shall decline going into the mathematical part of it here, as all mathematics advanced on the subject since the days of Dr. Lardner are absurd, and it would bother us terribly to demonstrate the matter satisfactorily in a work like this, which is to be purely descriptive and practical.

To return to the practical part of our subject, you will follow the air coursing through the mines from the mouth of the drift, to the point it leaves the mine at the top of the air-way, and gather up information as the air gathers noxious gases and clouds of powder smoke in passing along. The head boss and the inside boss will assist to work out the example of natural ventilation.

You enter the drift with the air-current, and at a short distance the head boss takes the lamp off his hat and holds it out into the air-current, to try its strength.

"Why, Bill, you have a brisk current of air here to-day; how do you account for it?"

"The air is colder outside than we have had it for some time, and the colder the air is, the faster it travels," says Bill, thinking this is an end of the subject; but it is only a beginning.

Harry, the head boss, stops and halts you. "Now, here's a chance for *you*, young fellow; let us know all about this matter before you advance a step farther."

You repeat Bill's answer, and tell him it is owing to the cold weather outside.

"The weather be blowed," says Harry, whose love of pneumatics will never drive him crazy.

"But it is a fact, nevertheless," you repeat: "and if you must have the principles explained, you must practise a little patience, so sit down on this oil chest and listen to them."

You tell your companions what has been told in the beginning of this chapter. Harry criticizes and discusses as many points as he pleases; but Bill becomes sullen and silent, evidently not disposed to encourage such unfamiliar topics. He looks like one among his superiors whom he does not wish to acknowledge as such. You explain how it is that the cold air has reversed the ventilating current by becoming the heaviest outside.

But as Harry is frank in his remarks, you are as candid in yours. You take your lamp and swing it in the current, and cannot help making a calculation. By custom, we can approximately find the quantity of air in circulation, by keeping a lamp oscillating a few seconds in the direction of the air-current. Now, in passing the lamp *with the current*, the flame must be kept *vertical*, and the space through which the lamp oscillates measured by a practised eye indicates the velocity of the air per second. Now by practice one may become sufficiently expert to catch up with a current of air moving six feet per second; but it is difficult to measure a current by such means moving much above this rate of speed. Your calculation made in the manner indicated, you say: "Gentlemen, we have not any air to boast of here. The current moves the flame over to an angle of 30°, more or less; but it is not moving with a greater velocity than twelve inches per second. Sixty feet per minute into seventy-two, the area of this gangway, will give a result of forty-three hundred and twenty cubic feet of air per minute. This does not amount to the name of ventilation in a mine whence four hundred tons are excavated daily.

"What has the amount of coal mined to do with air?" asks Bill, gruffly.

"Why every block of coal you break from the face of a working-place liberates a certain amount of gas of some sort. It is not often pure air that is found in the pores of the coal strata, and the crevices of a mine, I can tell you! Carbonic acid, sulphurous acid, and other gaseous mixtures, as well as the explosive gases, of which we will say nothing now, are found in inconvenient abundance in such mines as these. And I repeat that every block of coal you dig out of the coal seam, and smash up into pieces, lets out a portion of the gas it contains, be that gas what it will. When you stop mining the gas is stopped from forming."

"That reminds me of a circumstance which came under my observation once, and may be accounts for what has always been a mystery to me," says Harry, "and as we have an abundance of time, let us have an understanding all around. I once worked at a *fiery colliery* where the air was no better than it should be. They tried to ventilate it by an old fire-lamp set up on the top of the air-way; but the air would reverse in spite of [here Harry uses an expletive]. Well, somebody was getting half roasted every day. The boss was Dutch, and when he engaged men, would caution them about the fire damp: 'You musht take care von him or he vill roasht you, he ish von [Dutch expletives reflectively].' This is all the effort our friend made to save his men. He thought by doing this that he had bestowed on them an immense favor. But I am getting away from the main part of my theme. The place had acquired an infamous reputation. 'The slaughter-house' one would say, another 'an active volcano.' There was a strike of several months' duration, and the mine filled up with water; they had stopped the pumps. After matters were settled, the mine was pumped out, and from that day to this not a particle of gas has been found in the mine. Now, was this occasioned by a cessation of mining operations? or had the water by its pressure acted in such a manner as to force the gas into the coal, and wedge it up? or, if not, what the deuce became of the gas?"

Harry laughs as he asks the question, which by no means is easily answered, provided one does not suppose it to be anything but a genuine case.

"I do not know, Harry," you say; "but a little of both of your suppositions might have had something to do with it. Did no change take place after the strike?"

"Oh, yes! the Dutchman left and another fellow came and built a large furnace at the *bottom* of the air-way."

"And do you not conclude that this was the secret of the business?" you ask.

"I do no such thing," says Harry; "because the new boss said there was no gas in the coal, and ventilation or no ventilation, from that day to this the mine has been safe from explosive gases."

"This is certainly singular, at the least; but as a phenomenon it is not unaccountable in such mines as these lying on so steep an inclination and subjected to so many changes. You will find a large portion of a mine containing one kind of gas; for instance, the carbonic acid; another portion of the same mine, carburetted hydrogen. It might so happen that after the new boss came and built a larger fire in a more effective position, the ventilation would be so much improved as to carry off the small amount of gas generated which had given trouble on account of bad ventilation, and in such a manner as to cause any one but a scientific expert to detect its presence."

"A scientific expert be ———," interrupts Harry; "who the deuce can be more expert than the man who has been a lifetime in coal mines, and who has made it the business of his life to find out the real nature of gases and air!"

"Why, the fellow whom you have just presumed to know, or to have found out all about the gases of a mine, is the scientific expert I refer to," you tell Harry.

"But this boss was none of your bookworms, and he was the best man I have met with for *carrying* air."

"Oh, then!" you say, decisively, "this accounts for the disappearance of the gas. The fact is, there was so *little gas* given off in the days of the Dutch boss, that the improvement effected in the ventilation made after he left, kept it in such a state of dilution that its presence was never suspected. A number of people will tell you that there is no gas in a coal mine if it is not found in explosive mixtures and attested by the fireman's try lamp. And in the particular mine in question it is barely possible that the mine advanced into an area that yielded carbonic acid instead of carburetted hydrogen. And carbonic acid is a gas to which we have not paid one-half of the attention it is entitled to. It is not hunger but it is gas that puts the white into the countenances of the men working here at your colliery," you have the hardihood to add.

"The deuce!" roars Harry. "Bill, let us sponge up the gutter with this scientific rat. He is sneering at our ventilation. Why, look! there is wind enough to blow my light out. To think that the rascal charges us with using up poor devils of miners and giving them their pale faces, is more than I can stand." Harry stares in your face and whistles a stanza from Yankee Doodle. After a pause of two semibreves, he asks if anything more is to be added to the subject.

"Oh, yes," you say; "if you will give me the honor of your attention, we will pass on with this forty-three hundred and twenty cubic feet of air per minute, and see what it is doing and to which points it is directed. But we shall pass nothing we see worthy of criticism without doing it full justice, even the gutter or *water level* in which you propose to duck me; it would be much better for the horses' feet, and for the railroad track if it were one foot deeper than it is."

"You pretend, then," says Harry, mimicking a sneer, "that you do not take your own boots into consideration, you swab, you! You don't care to tramp in here amongst the mud! What a pity it is for you! Let us go, Bill, or the fellow will detain us a whole day."

You plod on through the mud until you come to No. 4 breast, and then you stop, and cry "halt!"

Harry turns sharply around, and asks "why the halt?"

"Simply this, we are at the fourth breast. Those breasts we have passed are all empty; their schutes and manways and headings are all open, as are the manways and upper headings inside of us connecting to the air-way. Now at the air-way we lose quite one-half of our forty-three hundred and twenty cubic feet of air per minute."

"There you make a mistake," chuckles Harry. "The bottoms of the manways have doors to prevent the air from going from the gangway, and those shut against the wind."

"My mistake is a very slight one," you say, "that it is at this point, instead of the other; it is evident where the best part of the air is lost. But this will make no difference, for a reason I shall point out before going a step farther, your leave obtained to do so."

"Oh, go the whole hog, by all means; don't spare us a bit; let us hear the worst of ourselves; we have lots of patience," says Harry, mockingly.

"Well, then, pay close attention. The upper heading through that pillar, dividing the air-way from the breast-room outside of it, is the only communicating passage between those breasts outside of the air-way and the air-way itself. The heading is three feet wide, by two high. This forms an area of six square feet. The velocity of the air through the heading to which I have reference, will be about six feet per second; quite a brisk current, and it carries twenty-one hundred and sixty cubic feet per minute. If this passage could be increased to half the size of the gangway, assuming that all its connections could also be similarly enlarged, or so that they made together a connection with this gangway of an aggregate area of thirty-six square feet, the circulation of air through that heading would be twelve to thirteen thousand cubic feet per minute. This would be the measure for the quantity of the wasted air. Now, if the conditions of the air-passages and connections were left unchanged inside of the air-way, the ventilation would still be the half of forty-three hundred and twenty cubic feet per minute. No matter how large the air-ways and their connections might be, presuming them to be in the aggregate as large as this gangway, the velocity of the current would not be any greater through the small heading opening through the pillar inside of the air-way than it would be through the supposed enlarged area through the pillar outside of the air-way. Now we will suppose the heading through the inside pillar of the air-way to be enlarged in a similar manner, what do you suppose would be the result, provided the connecting passages were also enlarged?"

"Why," chimes Harry, "by your own mode of showing, we might expect an increase in the circulation of air to correspond with the increase of the area of the passages, the velocity of current being the same. But would not the greater distance affect the ventilation of the inner portion of the mines?"

"Not to a very great degree, for the small difference of a few hundred yards. The distance affects the ventilating current inversely, as the square root of the whole is compared with the square root of a part; but the areas of passages affect them as are their respective proportionate areas to each other."

"None of your algebra or roots, if you please; fair, square *round* numbers we will listen to with all due patience; another hint at geometry, and we pass on," says Harry, shaking his head and pretending an air of grave determination. "But, Bill," he adds, turning to that worthy, who seems to take no interest in what is passing, "suppose we stop up all the openings of these breasts outside of No. 7, what will our friend say?"

"That you would not be acting wisely or accomplish that which you desire, if it is an increase of air to the inside places. The air passing from the outside breasts would simply cease, and the air in the inside breasts would remain the same as it is now, because the current, under existing circumstances, would not exceed *six feet per second* anywhere. Therefore, by stopping up those outer breast-rooms as you propose to do, you would effect no improvement!"

"Supposing, then, that if we made the air-way of only the size of *one* of the headings feeding it, what would be the effect generally?"

"You would reduce the total amount of air to one-half of what it is now. The velocity of air in the air-way being but six feet per second, would be supplied by the two currents coming out of the two headings at the rate of three feet per second instead of at six."

"In this case, supposing that we stopped up the outside heading, what would be the result?"

"The whole of the current would pass inside and return through the heading of the pillar inside of the air-way at the rate of six feet per second instead of three feet per second, as just now supposed."

"Now, I have got it through my numbskull as a RULE that a stated difference in temperature between the inside and outside of a mine, together with a difference in the levels of the intake air-ways and the discharge air-ways will give a certain velocity to the air, and if we wish a large quantity of air to ventilate our mines, we must have a passage or a series of passages having areas to correspond to our requirements. But, by George, how do we find out what are those requirements?"

"To find out this important matter, we will pass forward, if you please, and try to solve the problem before we return to the surface, by examining the source of the impure gases."

"How do you propose we shall go?"

"With this remaining half of the air-current."

"What! inside, and by way of the breasts up the air-way? What a punishment!"

"But the air-current will teach us a valuable lesson if we follow it up; you will see what a fog of powder smoke it takes upon its weak shoulders. Besides, No. 8 breast is finished and should be measured and 'squared up.'"

"Well, Bill, then we must go and measure No. 8 and give those fellows a chance to go and work in the next new breast to be 'turned off' the gangway."

Mr. William assents as if he were doing a favor instead of performing a duty to his employers. The measuring of the distances and spaces of a gangway, a schute, a heading, a breast-room, and the rest of the mining excavations, is usually performed by two of the "bosses."

It is the custom of the miner never to dispute the measurements of his work by them. It is a compliment he pretends to pay to their honesty. Now he knows right well that advantage is taken of him in very many instances; but for reasons politic he closes his eyes, or his mouth, if you will, in nine cases out of ten. They are not always the men who do the best work who get the best measurements. There is always a tendency on the part of some bosses to show their partialities, and the men who dispute with them in little matters, however just they may be, are not the men who receive the favors. But an honest miner will rather dispute concerning his rights than be the recipient of a favor he has no right to accept. And the poor fellow often comes to grief in consequence. We shall, for the sake of detailing the business of mining in its several branches in the best manner we can, enter into the disputes which so frequently occur between the boss and the miner which often lead to bad consequences in the mining business. And this is not at all departing from our descriptions of modes and methods of operating a colliery in the least; because the miner must be dealt with as well as the mines, and I have an idea that this very important item (the miner) of the mining interests, is entitled to a little more respect than a mining mule.

But you have left Harry and Bill standing in the muddy roadway of the gangway. They have a little dialogue of their own concerning the gutter on the low side used as the water level or drain. For the sake of information let us join them.

"Bill, would it not be better to have the gutter on the upper side of the track; cut away into the bottom slate for instance?"

"That is what I have never seen before," growls Bill; "it seems unnatural to have the gutter on the *high* side."

"But which is the high side?" asks Harry. "For my part, the gutter being placed in this manner in the bottom slate at the lowest point would be literally in the lowest side, and I would call it the low side. Here we have it," and Harry draws a sketch and discusses its merits (Fig. 9).

"Now here is your old gutter in the coal at *A*. Here is the gutter to be cut out of the bottom rock at *B*, and so deeply into it as to be used as a water course *after this lift has been abandoned* and worked off."

"What use would it be then?" asks Bill, bluntly.

"Don't you see it would collect all the water coming over the bottom slate from the surface and pass it off without bothering the pumps of our lower lifts, after the workings of the lower levels fall through or are driven into this water level gangway?"

Fig. 9.

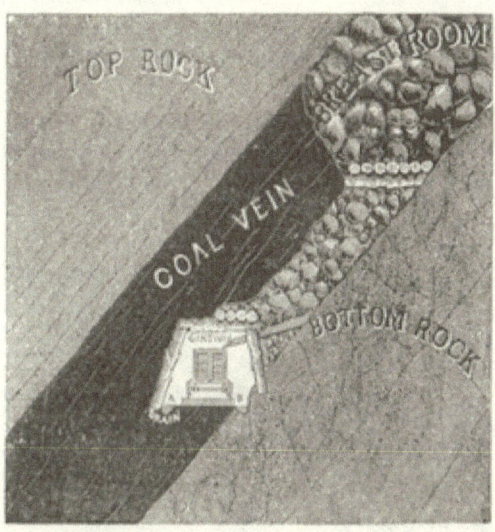

"But look at the extra expense of cutting the gutter in the rock; it would cost an additional two or three dollars per yard. And then what will it matter to us or our owners; we may be, God knows where, when the lower lifts are worked."

"Now, Bill, look here!" interrupts Harry. "If this were my property, and you were a lessee, you would not extract a pound of the coal from this water level unless you agreed to build a permanent water course as you progressed with your water level mining. I should not be so green as to lease any of my property to be ruined by any parties who would not lease with a clause inserted to bind them to such an arrangement. Look at that gutter! it is flooded. There are not less than three hundred gallons of water per minute delivered by that shabby little drain of yours. Now when this coal is worked below us—which it will be within a dozen years of this time—that water will be pumped up from the next lift, and then, after that, from the deeper lifts second and third, and Heaven only knows how far it will run down into the mines when they are opened below us in after years."

"Yes! but how would you build and support this permanent water level?" asks Bill, in a tone of assurance.

Fig. 10.

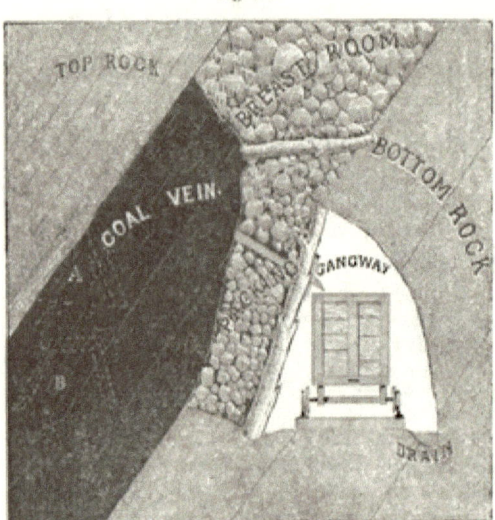

"By driving the whole gangway into the bottom rock, and by securing it with timber set up in the 'jugglar' fashion, according to Fig. 10. By using packing behind the legs in the manner represented, after a portion of the solid coal has been removed, we should avoid the serious crush which comes on timber which is wedged up against it with nothing intervening to mitigate the immense strain. Besides forming a permanent water level, it would form a permanent gangway, which is more desirable still, as it would not require relief timbers. Stumps of coal would not be required above the gangway. Then look at the room we have for a monkey gangway to use an air-course either in the position shown at A or at B (Fig. 10); a single prop would suffice at A and double timbers at B. In addition to these advantages we could work the breasts of our lower lifts clear up to this level and cause it to suffer no injury."

After Bill examines the sketches which Harry has drawn in his dirty time book, he ventures to ask concerning the cost of such a gangway.

"With the use of a compressed air drill," says Harry, "I could drive such a gangway for twenty dollars per lineal yard, which is less than this our present gangway will swallow up before it is finished, after it has been supplied with its sets of relief timbers, besides other repairs. In addition, we would be able to get about one hundred tons per yard more before we robbed any of the pillars."

"I do not see how we can do that," says Bill.

"I have just told you I would not leave the *stumps* above the gangway of this lift. I would drive my schutes to the top slate and open the breast close to the upper edge of the gangway, and thus have the benefit of twenty-four feet of breast more than I have. Then in the next lift below I would drive the breasts entirely up to this gangway and take advantage of the rock passage for an air-course."

"That would certainly be a great advantage to the people of the lower lifts. It would drain off the light gases they will have. Look, here are a number of gas bubbles escaping at all times in this place," says Bill, with a little more of grace in his tone. And he takes his lamp and puts the flame into contact with the escaping bubbles, and lo! a little blue flame shoots out of the gutter as the bubbles burst. "I bet they will have a hot shop down in that quarter of the mine when they come to open it out," chuckles Bill, in a manner indicating that he will not be of the party which is to contend against the gases in the lower levels of the mines.

"All right, Bill; I have just been fooling you into a better temper; we shall not drive the gangway into the bottom rock. The people here will think us crazy if we propose such a plan, and we are not the land-owners, nor are our employers anything but lessees of very short duration, unfortunately for themselves and the land-owners too!"

"We will go on and get to No. 8 and see if this scientific man has any suggestions to make about the air-current. He has told us not a word of the subject of drainage, nor of my proposed method of working a water level. I conclude they are subjects above his science," says Harry, in tones of pretended sarcasm.

"I can find not the shadow of a fault with the remarks you have made," you say. "Bill has suggested the great idea of using the road you have proposed to make as an *air-course*, which gives your water-course a double capacity. I have given you both great credit for your respective suggestions, and I shall take the first opportunity of sounding your praises."

Passing along the gangway, you are where the loaders are at work drawing the coal from the breasts. The air is getting thick with powder smoke.

"Look here, young fellow!" says Harry, turning to you. "What brings this powder smoke here? The air is going inwards and the breasts outside are all stopped up with coal; at least those are in which the men are blasting, and in spite of this, right here we have the smoke thick enough to be cut with a knife. How is this?"

You look at him unflinchingly, and tell him of the great power of absorption and diffusion of the gases, and of the incomplete batteries formed by heaps of coal, and the large areas of the schutes and other passages connecting the breasts with the gangway. And you tell him of the counter currents formed by air passing from large passages into small ones and then again from small ones into large ones, and show the principles and powers of endosmosis so well or so badly that he gives the matter up and passes on, leaving the air to drain through the masses of coal in the breasts and bring the powder smoke along with it as if it were an impossibility to prevent it.

Following up closely, you remark: "If we had the other half of the air current brought forward to this point it would play the mischief with this powder smoke. If you could make the headings three times the size you have them, you would give us three times as much air to carry it off. Why the mules cannot see the wagons they are bumping!"

"That is too strong, young fellow! we had better get inside, and I shall take you up the inside breast, and, by way of the headings, from one breast to another until I get you to No. 8, and then look out for squalls, if you have not the *wings cut down*."

CHAPTER XXII.

EXAMINATION OF BREASTS—MEASURING OF WORK—PARLEYS WITH THE MINERS.

So we pass on to the inside schute and climb up the inside manway, through which a tolerable current of air is blowing. (To understand how the manway and heading connections are made, see Plate XX.)

The inside schute has just been driven up, and at the top of it the breast has been set off or "headed over;" but the manway on the inside has not yet made a connection. Consequently the inside schute connecting with a manway driven with the breast just on the outside of it receives the current of air carried to this point, and thence the current may be traced from the heading above the schute to this manway; from this manway again to the upper heading; from the upper heading into the breast; and the headings and manways in the pillars intervening between this point and the air-way perform a like service until the air is delivered to the air-way, and passed off to the surface through the wooden shaft built on the top of it. It is away up on the mountain amongst the bushes, where, of a frosty morning, you will see a cloud of vapor caused by the condensation of the moisture and gases brought out of the mines. This cloud has a ghastly appearance, and gives out an unhealthy odor as it thrusts itself up into the pure air, which in cold weather seems to have a decided antipathy to the stuff, and apparently receives it as an unwelcome guest. But here is shown the contrast existing between the two atmospheres—that nasty gray cloud of the mine as if coming from the *infernos*, and that other beautiful azure of the heavens—the one breathing plague and pestilence and death, the other showering the blessings of the celestial regions over all creation.

For the benefit of those versed in systematic ventilation, we shall describe the manner of stopping the air more in detail.

The stoppings of the headings are *formed* by the masses of coal in the breast-rooms, filling them up as the excavation becomes filled with loose coal, and as this blocks up the entire chamber formed. But such stoppings do not seem to be air-tight in spite of the large masses of coal lying against them. Nevertheless, if a large amount of air were put into circulation through large passages and the currents were travelling at a moderate speed, the waste of air would not be appreciable. But when the reverse of this is the case the small leaks of air *soaking* through those masses of coal seriously affect the innermost places which are always the busiest and require the greater amount of air to ventilate them.

At the outside breast-rooms the air divides itself into two parts, one-half of which finds its way to the empty breast-rooms to pass at will among them to diffuse and carry off the gases generated in them, which are for the most part a mixture of the mine gases, that often predominating being sulphurous acid derived from the decomposition of the sulphurets of iron (Fe, S_2). This decomposition is accelerated by the action of moisture and the water which soaks into the rocks—top and bottom—both of which contain an abundance of those *pyrites*. If the rocks fracture and cave in, the pyrites are much more exposed to the action of the air and its moistures. This then forms the *white damp* of the miner, named so on account of the white top it puts over the flame of his lamp or candle, especially when it becomes mixed with fire-damp. It has also been termed *stone gas*. Its poisonous qualities are too well known to miners to need discussion just at this place.

Returning to the air-current and your companions, Harry and Bill, you will continue your course around the works, and chat a little with the miners in the breasts, whom you find to be of all the emigrating nationalities.

"Now, what do you think of this breeze," asks Harry, putting his light opposite the heading you have just crawled through into the inside breast.

"*The breeze* you refer to is well enough, if it were large enough. You see it is not projected with a greater velocity than one foot and a half per second. Now, one foot and a half per second into the sectional area of the heading, which is six feet, gives us five hundred and forty cubic feet per minute. This is the remainder of the total of the air, and that portion only that reaches the innermost workings."

The miners working in the inside breast have ceased work, and come down near the heading, and join in the conversation. Pat and Dennis work here. Pat says, "An' it's the terrible headache we get here before quitting time. It's a shower of red-hot prowther schmoke that comes into the breast through that same heading! Air! it's pison!"

"Shut up!" says Harry; "it's whiskey, man, that gets into that red head of yours. Bill, just examine Pat's cans, and see if he has not a pint of the cratur corked up there!"

"Arrah! now," chimes in Dennis, "it's just as much as we can do to get the bite we put into our mouths working in this powther schmoke. The coal's as hard as the divil, an phat betune driving the breast, blowing down top, driving the manway and holing the headings, we have our hearts broke! An' it's nine dollars a yard you give us; be the holy saints, its wurruth eighteen. The little drap we get doesn't settle the foul gas—"

"It's the *little drap* you hide in that five gallon demijohn you were hugging home the other evening you mean, I presume," adds Harry, interrupting.

"Faith, and that's the medicine I got for the old woman," retorts Dennis. "She's about being—"

"And has she been working in the bad air too?" asks Harry, without allowing Dennis time to complete his sentence, which promised to be interesting from a domestic point of view.

"You're the divil shure," joins in Pat.

"Now, a little business, if you please, gentlemen," says Harry, taking out his greasy time book. "Your top is not broken through yet, and I cannot measure your work for last month. The coal is hard, and your breast is in fair condition, and as I have no doubt you have worked honestly, I shall turn in a certain amount of the whole distance you have driven, and expect you will break through the benches before the end of next week. Next month I shall turn in the extra amount due for breaking through the seam. Harry has written a memoranda in his time book, and now puts it in his pocket, as if he has finished the business of the breast for the past month. Pat and Dennis consult each other with an expressive look that both understand as satisfactory; and Harry looks questioningly in the face of each, and an understanding is arrived at in less time than we can express it.

"Well! how is it, boys, ?" he asks, as he rises and prepares to cross the breast to the heading through the outside pillar leading into the manway, and thence to the breast driven on the outside of it.

"It's all right, sur, and much obliged to you for the same," says Pat.

"Well, good-day, and take care of yourselves, and don't forget to keep those wings squared up; if you leave them project, your top benches will not get down easily. And, Dennis, take care of the old woman."

"Go to the divel!" roars Dennis, and everybody laughs, and you crawl through the headings and manway and come into the next breast outside.

Here you meet with Jacob and Peter; Jacob Steiner and Peter Dundleberg. They are both drilling holes, and they are both smoking pipes. When you make your entrance into the breast they continue to drill and to smoke without any perceptible interruption or alteration being made in the churning of their drills.

Jacob is on the innermost pillar and Peter is on the *outside* corner. You first meet with Jacob, and Harry salutes him. "Wie gehts" is answered with Jacob's "Ziemlich gut." And the drilling—a slow, lifeless kind of a stroke being used, having neither vigor nor energy about it—continues, as it is being carried on, in a cloud made up of powder smoke and tobacco smoke. There is no steaming perspiration in the cloud of smoke; the strokes of the drills are neither sufficiently rapid nor vigorous enough to extract a drop of sweat.

"Let us go, Bill," says Harry, looking around at the top coal and pillars as well as his lamp will allow him to do through the smoke in the breast.

"No! we should make those men blow down those wings," says Bill.

"*You* must attend to that, Bill. I shall not bother with them. We can keep back enough of their pay to warrant the work being properly executed. See that those wings are taken off before pay day. Guten tage, Jacob!" says Harry, as he enters the outside heading. "Guten tage," grunts Jacob, as you leave them to their drills and their pipes and their smoke.

When fairly in the next breast, Harry sits down on the loose coal near the face where two miners are at work in a lather of sweat. You seat yourselves close to Harry. The miners stop work and come and join you. Harry looks up at the top coal and asks, "How is your top coal, boys?"

"It's good for half a dozen kegs of powder, by George!" answered one of the miner boys.

"Say, Jack, he ought to allow us about four or five kegs to help that top down;

it's just as solid as a bell, and it will make the kind o' stuff that 'll put a white hat on the owner; they can well afford it if they choose to."

"No, sir-ee!" says Harry, promptly. "Men who can afford to lose fifty dollars a month on a race and the like sports can buy their own powder. It's not only the money you lose but lots of your time is lost as well. Picnics, races, walking matches, 'go-as-you-please,' and the like. The top coal did not trouble you a week to-day. Charley, you are a bad boy!" Harry shakes his head, disapprovingly.

Charley is silent, and does not relish the laugh the others are enjoying at his expense. A week to-day has reference to an affair of Charley's which ended in his being cleaned out by a couple of New Yorkers in some of the New York gambling dens, where he had had the hardihood to venture with a few months' savings.

The banter stops at this allusion, and Harry lets out on the boys in this wise: "Now, lads, you are very foolish. You are now working like horses to make up for the time and money you have lost at gambling lately. You are both alike; not a toss-up in the difference between you."

"But we have reformed," says Jack.

"Ah! yes, you have reformed, no doubt. You have pawned yourselves. For the next two months at least your pay is pledged."

"And by ——, sir, we are the boys who intend to keep *that* pledge," says Jack, "if we cannot keep any other."

"You are all right there," says Harry. "No danger of your going back on the friends who pull you out of the mire you stick in so often. Reformation! Oh, certainly, till you get another ten dollar bill in your possession." Harry looks knowingly into the boys' faces, and says, "Let us leave these lads to their good resolutions which are just as well, if they are not of a voluntary character. Take care of yourselves, and don't let your hurry to catch up lead you into any danger. There's more accidents to be scored to this hurry-up never thinking system of working in mines than to any other I know of. Good-day, boys!"

Before you get into the heading leading to another breast, "the boys" are again working with a will.

"Aye," mutters Harry, as he is crawling through the heading; "they're like a great many others who work like horses, and spend their earnings like asses."

"You better hold up there," calls out Bill, who is following Harry through the heading; "I think some one is working in the manway." Harry stops before he reaches the manway, and calls out, "Halloo up there!"

There is no response; possibly owing to the sound not reaching the men in the breast on the outside of you.

"There's no one up here, Bill," says Harry.

"You had better wait a little; I heard some one crawl through the upper heading, or my ears have played me false. Sing out again to make sure," advises Bill.

There is no need to do so; in a second more there was an explosion. A blast is fired in the manway, and the wind driven from it blows out your lights at once. At the same time the coal from the blast shoots down the manway past the end of the heading in which you sit waiting.

"That's a near shave, Bill. But for your timely warning, boy, some of us would have been toppling over and over, like the coal rushing pellmell down to the platform. I'll give those two Welshmen particular, when I get up to them!" Harry puts such an emphasis into his threat that one could not but guess what was meant by the "particular" which was to be presented to the Welshmen.

You crawl into the Welshmen's breast without further delay. You find them "taking their whiffs" near the breast.

"Why don't you men give notice before you fire a shot!" asks Harry, in a tone of anger which made Billy Jones and Billy Williams (usually called the two Billies) take their pipes out of their mouths, and stare at you as if you were ghosts come to them without your lights burning. They exchange looks with each other, and with each of you, as if they did not comprehend exactly what was really meant. "Has anything gone wrong, man-dear?" asks Billy Jones in a tone of consternation.

"My good Lord, what be the matter, men? For good—ness sake do speak, some of you!" implores Billy Williams.

"You know you have just fired a shot in the manway!" says Harry, severely.

"Yes! goodness gracious, yes!" chime in Billy Jones and Billy Williams at the same time, as if it were a chorus they were repeating (they were noted singers), and they waited in breathless expectation for what was to follow.

"Don't you know what is the consequence when you fire a shot on the top of people coming up the manway?" asks Harry, beginning to enjoy their excitement and confusion; and to punish them he delays the explanation.

"Who is hurt?" asks Billy Jones in alarm.

"My goodness, be anybody killed?" implores Billy Williams, looking searchingly into Harry's severe countenance.

This is more than you can bear, and there is a roar of laughter all around.

Jack and Charley had heard the firing of the blast, and thinking it might have caught some of you in the manway, have come to ascertain for themselves if your party has escaped. Their appearance does not in any way pacify the excitement of the two Billies. Jack and Charley, seeing you are all right, and guessing the drift of Harry's sarcasm, are ready to join in the merriment.

"Well, this is better than we hoped to find things here when we came," says Jack, "and I'm heartily glad that we have not to use a stretcher for any of you. This is worth a pull. We'll have a whiff, at your expense; so hand over your baccay fobs."

The Billies produce their pouches well filled with the bird's-eye.

At least half an hour is spent with the two Billies, and it is taken advantage of to call to mind some of the blasting accidents, and the very near escapes that each of you have known of.

Jack and Charley leave, and while still in the heading, Jack calls out sarcastically, "Say, Billy, you'll give us a knock when you are about to blow a hole through into us, won't you? The fourth of July 'll be here one of these days; let us have a chance of spending it."

"You bet we'll knock you a signal the next shot we fire in the manway; we'll not forget this lesson, you may depend," responds Billy Jones.

"And thank the good Lord it is no worse than it is, responds Billy Williams, piously; "I can sleep this night in peace."

You leave the two Billies, and having spent considerable time so far, do not delay. No. 8 breast measured finally, leaves you at liberty to gather up your tools, and take them to the surface by way of the air-way, through which your late companions have just crawled. There you experience what you have often done before, the effects of a sudden transition from the atmosphere of a mine, highly charged with deleterious gases, to the pure atmosphere of the surface, as it is found under the shade of the trees. You respire it with your lungs fully inflated, as if desirous of making good any loss you have suffered by your incarceration in the mine. You mournfully reflect on that system of ventilation which deals out a substance so sparingly to the miner, as if it were a scarce and costly article, instead of being the most voluminous substance in the universe, and one which will *thrust* itself into every remote corner of a mine if you only allow it to do so, and provide a simple road for it to travel.

CHAPTER XXIII.

GENERAL CONCLUSIONS, WITH A COMPARISON OF THE DIFFERENT SYSTEMS OF MINING.

It is the usual practice of writers on coal mining to divide the work off under several systems, and to dwell elaborately on each. In the preceding pages we have given but two examples in mining; but these have been amply illustrated and described, and comprise the chief features of practical mining.

The first section speaks in a general way of mining, and of the system of long wall, which varies very little from working out the broken by the board and pillar system which we have described in the second section. The one takes out the coal advancing, and cuts its roads through the stratum forming the natural roof of the coal seam as it sinks into the excavation. The other system takes its roads to the boundaries, and runs a network of passages or excavations through the coal bed to form what are termed the pillars. These are finally taken out, and the roof above them falls into the spaces which those pillars have occupied.

Generally, those pillars are first taken out at the mine's limits, and the space left between those limits and the pillars being operated on, forms what has been termed a goaf, through which no roads are required.

There is not so much difference between getting out the coal of a seam by either of those methods as there is between the coal seams themselves. One seam may be thick and highly inclined; another may be thin, and lying nearly horizontally, or the reverse may be the case. They require to be differently treated in some respects; but the main object to be kept in view, is, or should be, to take out the whole bed in a *property* with the least possible delay and expense. If we lay off a thick seam of coal lying at a high degree of inclination, which is to be worked in pillars, ten to one that not one of those pillars can ever be taken out, except under the most favorable conditions. Pillars they tell us have been *run out* or crushed out, and their products have been loaded on the gangway, as they have run down on the floor of the seam. But this is as it may be, and if in such cases the rock comes down after a portion of the pillar has been run out, this stops the further produce of the pillar. This is according to the general rule in the United States in the thick beds of anthracite which lie highly inclined, and how this is so, any one studying the plates which show the mining of such seams in detail will readily understand. Plate XX. shows us a range of pillars as they are left in a mine which is termed "worked out" on the map of it, or on the map of this particular portion of it, and more than one-half of the coal the seam originally contained is still left in that particular mine.

The system of working coal in France by *remblais* seems to be the one best suited to all thick and highly inclining coal beds. But its introduction into this country would be attended by much opposition. Miners do not relish the idea of changing from old-established plans to new ones; and as those miners have a right to question the merits of any proposed change of this kind, on account of the great risk they, in all cases, run, they cannot be blamed for raising any reasonable objections, which few would like to take the responsibility of ruling out.

Yet it is on account of its safety that the French mode of working can be most urgently advocated as the plan to be used in highly inclined and thick beds of coal. The lifts being kept under two metres (about six feet six inches), keeps the roof within the workman's reach, so that it may be easily propped and kept secure until it is packed up by the remblais. Besides, the extent of roof over an advancing face being so limited, no serious difficulty can be experienced from the effects of a crush, as if it were the solid rock pressing on the coal cut away from the seam.

The board and pillar system, as applied in the Yorkshire mines of half a century ago, very nearly approaches long wall, because not more than one-fourth of the coal was left in pillars. These latter the roof crushed out as it settled down, and their contents were lost.

The same plan is pursued in the Wyoming Valley; and on account of thin pillars being left when the roof breaks extensively in any one locality so as to weaken the lateral support, a "cave-in" is the consequence, which is often very extensive in character, on account of the pillars being so thin as to allow them to be crushed down by the increasing weight of the lowering strata. It is to be feared that grave results will be the consequence, if such caves-in take place near the Susquehanna River, which flows through that valley.

But the plan of mining pursued at a colliery is not of any greater moment than that of arranging the men, and dividing the work. It is as much owing to the admirable system of dividing the work among the men that such great results have been attained in the Newcastle coal field. From some of the coal seams worked there under four feet in thickness, two thousand tons of coal are sent daily to the surface.

A miner there has never to put down his own tools with which he is well practised to take up those of the deputy, to set his timber or lay his track, or put up his brattice. When he goes into his board-room or jud, he expects that the wagons will be brought to the end of his track as fast as he needs them, and that there will be little or no interruption of the work while his short shift lasts. He is seldom disappointed in this matter. He does not often get a chance to sit down, and it is well that he does not, for then the sweat would dry upon his person, and he would become chilled; which, to be repeated day after day, as it is in some localities among certain miners, would in time produce miner's consumption, a disease as prevalent in our mines as it is exceptional in the mines in the north of England, where the hours of the miner are as busy as they are short, being little more than six hours, during which the pick is quickly and skilfully plied. Take away the mining engineers of the north of England, and the hands trained to work systematically, and put Thomas, Richard, and Henry in their places, and they will do no better than they do in other parts, where little attention is paid to such an important matter as the proper distribution of labor, which in those mines is as well regulated as it is in the establishment of the Baldwin Locomotive Works of Philadelphia, which is a model one of its kind in this important particular.

But credit is not due to the men any more than to the "lads" of those mines. The trapper at his door is hardly initiated before he is taught something of the post to which he is to succeed next, which usually is that of the driver. At about twelve years a boy is supposed to be sharp enough to manage a horse well trained to perform the peculiar work he is required to do. He waxes in this position of driver, and then falls into the chance or irregular one of timber leader, or station clerk, or something else, until he is able to *put* with a pony, or by hand. This carries him into manhood, with generally a great deal of brass in his face, and sometimes enough

GENERAL CONCLUSIONS.

of iron in his *heart* to carry him safely through his future existence. He may take his place next as a hewer. If he turn out well, he may get to be a deputy, and then an overman; and in time, if he have any one to direct his studies, may become an expert at mining engineering. But those of the pitman who have risen to this stage have been rare exceptions. However, Stephenson, the engineer, and Hutton, the mathematician, were both pit lads, and worked in the mines of the Newcastle coal field.

So far we have made little or no allusion to any part of the machinery of the mine, beyond that used in underground transportations, not even to the *on-setting* machines, now becoming introduced into the English mines, nor to the self-dumping arrangements which, properly speaking, hardly belong to a work whose main object is to describe the manner of the mining out of coal beds.

Our plates have been made with the object of giving general and detailed ideas concerning the mining of coal and the ventilation of the working faces, which generally give out gas in a certain ratio to the *amount of coal cut in a given time*, more than in any proportion to the area or extent of mine opened. The Plates I., II., III., and IV. have shown how the ventilation is conducted and distributed around the mine. Plate III. shows the crossings and mode of "splitting" and regulating on a large scale. Plate IV. shows not only an officer who has his eyes and ears at all times applied to detect variations of the air-currents, etc., but also other important details whose uses are manifestly obvious to any observer.

The succeeding plates, up to IX., belong as much to the transportation of the mine's produce as to the cutting of the coal, and illustrate many details which are not generally understood, even among miners, in many parts of the world, where mining regulations have not been reduced to a general system.

Following up in the line of our illustrations, we next come to a series of views which represent the working out of the coal finally; and when this is done by the system of juds or lifts as the mining is retreating, or by long wall in advancing, the actual difference existing between the two methods, as far as the digging out of the coal is concerned, does not really amount to anything serious. The forming of the gate-ways is the chief feature in which the difference consists; and perhaps, too, because the holing and advancing are done in the line of the advancing wall face, instead of being taken off in "lifts" or "juds," as in the case of working the pillars in the *broken* workings.

A change from the wooden to the iron prop is also peculiar to long wall in propping. This causes the roof to break off in lines parallel to the props as they are set in the wall face in rows; and this brings the falling of the roof to some extent within the control of the mine officers.

In either case, the miner takes out the coal, and allows the roof to cave in behind him, and as far as the management of the roof is concerned, he has nothing to do with it.

Particular attention paid to the propping up of the roof is of the utmost importance everywhere; and several views have shown us how carefully this is done in the mines of the north of England, where the deputies attend to such matters. Where the miners do this work, as they do in some other localities, the propping up of the roof is often sadly neglected, and the men prefer to run great risks in order to avoid a little trouble.

Following up we transfer our scenes from works in mines lying nearly horizontal, to those lying on a dip of about 45°, and by a series of plates have given views of the different excavations cut in the beds of anthracite found in Pennsylvania, United States of America. Plate XX. has given us a general idea of the plan of working which seems to be universal in those mines. As the pillars are left as a rule, in spite of the pretence made to rob them, much of the coal is left in the mines in such an isolated condition as to render it an impossibility for it to be approached by any other means than by cutting a road through the bottom rock, in the manner shown in Chapter XXI., and no doubt many of the mines will be worked over a second time by the mode we have suggested; because, the bottom rocks being in most cases hard, are not likely to be affected by the crush which is brought to bear on the gangways after the pillars have been disturbed.

By late surveys it has been found that little over one-third of the coal has been sent to market at the first mining over of the various properties, which are termed on the maps as being "worked out." Reference to Plate XX. should explain how this is, and Plate XXIII., detailing important features, assists to show how the immense bodies of coal left in pillars swell the amount of deficiencies, when the

GENERAL CONCLUSIONS. 147

area of a mine worked over is called to account concerning its original contents, in order to compare them with its actual product.

The breasts of coal are not always driven up the incline of the seam, as we have shown them to be; but they are so similarly disposed towards each other when they are driven level or diagonally across their floors, in order to ease the grade of the railroad track used within the excavations, as not to alter the system of mining used in the least; it only changes the direction of the breasts.

With a few exceptions of splitting or dividing the main body of air into several currents, the ventilation is effected by the air being coursed around the works in an undivided body; although at most of the collieries large fans are at work, exhausting the air from the mines, whose force in a number of instances is spent in a bad underground arrangement of the air-courses.

On account of the great thickness of the seams, good opportunities offer themselves in favor of ventilation; because, in some mines, large air-courses may be driven at less expense than in thin seams. This is especially the case where the seams lie at a low degree of inclination.

SECTION IV.

THE VENTILATING FAN—UNDERGROUND FIRES—ELLENVILLE COLLIERY,
MAHANOY COAL BASIN, PENNA.

CHAPTER XXIV.

THE VENTILATING FAN—HOW IT SHOULD BE CONSTRUCTED AND ARRANGED—PRINCIPLES
OF ITS ACTION DESCRIBED.

To promote a current of air in mines, there are two modes of destroying the equilibrium of the atmosphere, a condition which must be accomplished before any movement of air can take place. By the use of a large furnace, or a set of large furnaces, such as are used at some of the large collieries in the north of England, a powerful current of air may be put in motion. As many as two hundred thousand cubic feet of air per minute have been used to ventilate some of those mines. By weight this amount of air is over seven tons. This, in one remarkable instance, has been accomplished by the use of three furnaces, acting in combination. But in this one instance, at least, the aggregate grate surface was nearly four hundred superficial feet, and the consumption of coal approximated a ton an hour! This is an expensive method of raising the wind, even if the coal is consumed at Newcastle!

The action of the ventilating furnace is very generally understood, and its power depends on the amount of heat invested in a body of air ascending in a vertical shaft above the level of the furnace. The deeper the shaft, the more powerful is the furnace in which it is placed as a ventilating power. The furnace was the first step made in the direction of artificial ventilation, and its use has been dwelt on at length in other works. It depends for its action on those principles set forth in a previous chapter, giving an example of natural ventilation, and it is not necessary to repeat them here. This furnace system of promoting artificial currents is fast giving place to another in the shape of the steam fan, and, in a few cases, the air-pump; the latter not proving as suitable as the former, perhaps on account of its complications, and of the friction and weight of the numerous valves which require a certain force to move them at the changing of each stroke, and which are not kept in order, except by close attention and considerable repairs. For these reasons, the rotary fan, which is the most simple of our ventilators, will take a prominent place in the ventilation of mines.

As soon as the merits of the two kinds of ventilating fans have fully established themselves, it will be found that the rotary fan, which projects its air by the so-called centrifugal force, will prove the most applicable and most effective. The other, termed the screw-fan, can never be made to project the same amount of air as the above, because the speed of the screw is so much less than the speed of the vanes of the other at the point of its great velocity, on which it depends for its projectile force, that the amount of air put in motion must be considerably less in the screw fan, when it is of the same diameter and running with the same speed as the common rotary, or if you prefer, the *centrifugal* fan. We are assuming that the best examples of the fans are understood in the above comparison. For the reasons just shown, we shall confine our further description to the fan as it should be made, to produce the greatest possible effect with any wheel of given dimensions.

The fan, properly speaking, is a projectile machine, and the particles of air on which it operates are the bodies projected.

The outer extremities of the revolving vanes form the points which determine the projectile force of a fanner or fan-wheel; because these being the parts of the machine which move with the greatest velocity *drawing* the air *after* them at the *same rate of speed*, project them into the atmosphere beyond, at very nearly the same rate of speed, provided the air have free access to the wheel, which in most cases it has not, on account of the faulty construction of the fan cases and connecting passages more than on the faulty construction of the wheel itself.

Any flat surface moving through the air is better than any other for drawing *after it* a body of air. Moving in a straight line, the said flat surface will push a limited amount of air in advance of it, which will cushion itself against the front surface. This cushion will form itself more like a cone than a sphere in case of a round disk, and more like a pyramid than a wedge, in case of a parallelogram. This air will push off the air in advance in a manner slightly diagonal, and the displacement is made more rapidly in front than in the rear of such body; after which, the air will flow in a column of considerable length.

A fan wheel operates best when its fanners are set in the same plane, with a line running through the centre of the revolving shaft, being secured properly to arms projected at right angles to it. To obtain the best possible results from such a wheel, which is to take air on both sides, the width of the vanes should be equal to the radius, and the width may be slightly increased at its outer extremity. The vanes may run down to the shaft, according to the taste of the constructor; but the arms or centre to which those vanes are secured should be made in such a manner as to offer no surfaces to act against the ingress of the air to the wheel, and the vanes should be made without any beading or border, which will prevent the air from entering into the circular space, through which the fanners revolve, or to escape from it as it is projected by the edge of the fanners.

For instance, if a lip were bent over at right angles, and formed on the outer edge of the fan, on the backside of it especially, it would act in the same way to retard the air from escaping, as a shovel would hinder the material from being thrown away from it, were an edge or lip turned up on its front border, and a certain portion of the fanner would be rendered ineffectual, which would be in direct proportion to the area of the lip so formed.

If this is true concerning the outer edges of the vanes of a fan, it is also true of the sides of them, past which the air must enter to the fan space. Any lip or other projection formed on the side extremities, tends to reduce the area of the inlet, and oppose the in-going current in a ratio equal to their aggregate areas. Nor does the setting of the vanes so as to incline or be curved back in a direction contrary to their movement effect any useful purpose whatever; such measures tend to add to the friction, by adding to the weight of the machine, and thus become the means of expending a certain amount of power uselessly.

For purposes of illustration, we shall give shape and dimensions to our fan-wheel, and put it in motion, to explain its mode of action.

The vans are five feet wide, and run down to the shaft, or rather to the hub fixed on the shaft to which they are secured. The wheel is ten feet between the extremities of the two opposite vanes, or in other words, ten feet in diameter. The area of the circle in which the fan revolves is 78.54 feet, which, multiplied by 2, gives us a product of the two circles (that is, one on each side of the wheel), equal to 157.08 feet. Then the width of the vane being five feet, and the circumference of the circle 31.416, gives us as their product, 157.08, which is the area of that annular space opening to the fan from the outside. Now, as it will be shown that through this annular space the air is projected by the action of the fan, it is here shown that this outlet exactly corresponds to the combined inlets of the two sides, and that the velocity of the air into the sides of the fan will be exactly the same as it is through its annular outlet.

If such a wheel were put in motion in the open air, and measures were applied to determine the direction of the air, it would be found that at all points in the circumference, the air would be projected in an infinite series of lines tangent to the circle, which the vanes describe at their outer extremities, and, also, that the air in motion within the circle proper would by contact project the air lying within a certain distance on each side of the vanes at their outer extremities, so that the mass of air moved and operated on would be greatly in excess of that contained within the limits of the circle described by the revolving wheel.

In the open air the velocity of the air-currents promoted will be very nearly equal to the velocity of the fanners at their outer extremities, as has been fully shown by a series of experiments made with a wheel eight feet in diameter, wings two feet wide, with the inlet open to the outer extremities of wings only on one side. Power up to ninety horses was used to determine the effects of such a wheel, and the amount of power required to attain certain results.

The experiments alluded to were tried in Philadelphia by private parties for private purposes in the year 1872, and it was also found that, for high velocities, two vanes answered as well the purpose of projecting air as four did. No pressure above five and one-half inches of the water gauge was obtained, and the air-current then moved at a velocity of seven thousand feet per minute, or about one hundred and sixteen feet a second, which, during part of the trials, attained 120 feet per second. This current raised short pieces of yellow pine board in the air, and projected them to a distance of thirty feet from the wheel.

Returning to our fan example, if we wish to reduce it to a practical application, we must either incase the wheel, so as to collect the air projected from its outer portion, and conduct it to any desired point, or cause those currents coming from some remote points towards the fan, to course through whichever passages we desire them to do, and then finally to take them into the fan through trunks opening into its sides. Those trunks and passages should never be less in their aggregate area than that of the combined inlets to the wheel, and in all cases an advantage will accrue from making them about one-fourth larger.

In the event of an external casing being used, in which the air is collected, to be driven away in a body through the passages with which the casing connects, and under a pressure something in excess of the atmospheric pressure, a fan acting under such conditions is termed a blowing fan.

If the air is drawn through a mine or series of passages, and conducted to the sides of a fan, the air in those passages is less in pressure than the atmospheric air, which on this account forces fresh supplies into the passages, as these are being exhausted by the working of the fan. This is termed an exhausting fan, and it is mostly by exhaustion of the air that ventilation is effected in coal mines.

As the air is projected from the fan at all points of the circle alike, it is evident that if we put any case or obstruction in the way of those projected particles, we neutralize, to some extent, the effect of the fan. Thus, were we to continue a casing around the wheel and quite close to it, there would be no action and no air-current; the air in the wheel would simply revolve. The same thing would take place were we to close up the sides, to prevent the flow of air to the sides of the wheel. The air in the wheel in this case would simply revolve in a rarefied condition, the extent of rarefaction depending on the velocity of the wheel; and this, because no fresh supplies could be obtained from the atmosphere on account of the closed sides. Experiments have shown, therefore, that in order to get the best results from wheels of any given size, the whole of the sides (in case of a double fan), as well as the annular outlet, should be left unobstructed, and that no part of an outside casing, collecting the projected air, should be nearer to the vanes than the radius of the circle described by their extremities while revolving.

As for the sides, they should be left fully open, the edges of the trunks being carried to within a few inches of the extremities of the wings, when the fan is of the exhausting variety.

No exhausting funnel, like the one now used so extensively, should be applied, for it effects no useful purpose, and it is only used to gratify the whims of those who apply it. If anything of this kind is necessary at the outlet of a fan, it should be made in the shape of a *wind* shield to protect the projected air-current from the action of a strong wind; for, should a strong wind blow against a current of air coming from a fan, it is obvious that the one will be, to a certain extent, counteracted by the other.

No less a person than Mr. Warington Smyth, inspector-general of the English mines, doubts the utility of the expanding funnel and fan case, both of which, there is no doubt, will in time be dispensed with, when the action of the fan has been more fully investigated and more satisfactorily explained. It appears that it has been introduced with the Guibal fan, which is a machine, as far as we know, that does not do any better than any of our commonest and much less costly wheels.

Confining ourselves more strictly to the *manner* of fan action, we will suppose a wheel in rapid motion, which has four vanes. The air in front of the vanes does not cushion against its flat surface in the same manner that we have shown it to do, in the case of a flat body moving through the air in a right line. But the air behind the vane follows up at the same rate of speed as the moving vane itself. As direct forces act in right lines, the motion of the air at the extremity of the vanes is in a right line, or has a tendency to move in a right line, which is *tangent* to the circle described by the extremity of the revolving vanes. The particles of air leaving the vane draw on all the particles or affect all the particles forming the body within that space, bound by the two adjacent vanes; and the entire mass of air, in fact, within the entire space in which the fan revolves, is in a like manner affected, and possesses the same amount of rarefaction. It would be impossible to conceive this to be otherwise, for if we put in rapid motion a body of fluid of any kind through a body of the same fluid at rest, it will be clear that a portion of the body at rest will be so influenced by the action of the moving body as to take a motion in the same direction. Now, as the fan action is as much due to the fluid in which it works, as to the vanes for a communication of motion to the particles at rest, it can readily be conceived what will be the effect of the air on the outskirts of the vanes, when there is no point of the circle through which a vane does not pass several times per second, and from which the air is not radiating at all points of the circle, in lines tangent to those points, as if it were a mass of fluid revolving, and incessantly bursting asunder.

You can well imagine what would be the effect of a mass of any light substance, such as flour, for instance, when ground to powder or meal, and placed between a pair of disks, and clamped just enough to retain the mass between their faces, and then caused to revolve with a high velocity. The mass would be driven out, as if by an explosion. There would be no point in the circumference of the disks that would favor the expulsion of the mass more in one place than in another. In case of flour being used to illustrate fan action in this way, it would be shown to form a cloud of dust in the air, of which the disk would remain as the centre.

It is about the same with fan action in the air, which, in case of pure air, does its work in an invisible manner. If a mass of smoke caused by the explosion of a body of powder or otherwise, were injected into the air-current going to a fan built according to the manner we have shown, and also in the same proportions, so that the air would not be revolved uselessly in the wheel, the action of the fan could be made to explain itself to any common-sense observer.

If the fan is at rest, the mass of air contained in the space through which it revolves, will represent a cylindrical body, very similar to that body of flour that we have supposed to be held between the two disks. But there being in this instance an unlimited supply of air ready at hand to supply the place of that thrown out by the revolving of this cylinder of air, the supply and projected currents resolve themselves into two continuous streams. But in some establishments grain and sawdust are moved through boxes which connect with revolving fans. Then it will be seen that the form of vane best adapted to *draw after it a body of air* to start it in motion, is a flat one, moving at right angles, as nearly as possible, to the line of its direction, and when made of any other form, it is because of people's whims or false *opinions* concerning the principles of fan action.

As for the proportioning of horse power necessary to work air-currents through fans and through mines, no very definite conclusions have been arrived at; and theoretical speculations and practical results have been so much at variance with each other, that it is not safe to offer rules for the guidance of those about to erect fans, other than those which have been deduced from actual practice. The same kind of problem is to be dealt with in this case as that which governs the movements of floating bodies in water. A steamship, for instance, is a fair example of a floating body which, when moving at a high speed through the water, turns Dr. Lardner's theory—beautiful and correct in every particular as far as it goes—upside down in its practical application. This is owing to some practical advantage derived from the force of momentum, as that force accumulated in the steamer by the action of her engines at the commencement of her voyage, which is never lost until the steamer is again brought to a state of rest. As it requires some time for a steamer to be got fully under way by the working of her engines at full speed, so it requires a long time to put in full motion a body of air in a mine, which will in many cases represent an incredible amount of momentum, if moved with a high rate of speed through the passages of an extensive mine. This momentum will be such as to continue the current of air in motion for an indefinite time after the stoppage of the ventilating force giving it its primitive movement. Taking all things into consideration, we are liable to underestimate our ventilating forces, and by calculations made according to rules set forth, founded on the laws of force and resistance as we are taught them, we would err, and be very wide of practical results on account, perhaps, of not being able to find out what is due to the momentum of the air-currents themselves, after they have been once started in motion.

As in the case of marine engineering, the power required to move certain masses of air with certain velocities within given times, must be determined by a comparison of actual results.

The power necessary to work a ventilating fan is that which is required to move the air-current with the required velocity, *plus the friction of the machine*. No matter what description of fan or other machine is used to give motion to the air-current, the power necessary to give the air-current motion will be the same in all cases. For this reason, one fan or other machine will be superior to another, only when it does not consume so much power uselessly by friction. For example, if a fan is used to promote a current of air that is eight times heavier than another, that is doing the same amount of work, it will waste eight times as much power in friction. When fans are cased up in the manner in which we so often find them, with the inlets too much contracted to prevent the air from entering freely to the wheel and the outlets closed and contracted to about one-fourth of what they should be, as in the case of the Guibal fan, and some of its imitations, to do the same amount of work that a fan should do, if constructed strictly in accordance with scientific principles, a machine about eight times as heavy is required. Now, allowing a wheel like this to promote the ventilating current with the same force, the friction it causes will be eight times as much as that caused by a machine doing the same work, whose weight is only the one-eighth part of it. That fans are made so much at variance with scientific principles is really astonishing; but so far, the construction of those machines has been left more in the hands of the manufacturer, who has followed too closely in detail the absurdities of the early machines.

For purposes of mine ventilation, where such immense volumes of air are necessary, the lightest and best machines should be used in all cases.

To prevent friction, the fan-wheel should be made light, and the diameter of the journals and shaft as small as possible, to be consistent with safety. The journals should be long—not less than three of their diameters—to prevent wear and tear, and of steel, to give rigidity; and the bearings should be set on a solid cast-iron frame, and kept in perfect line with each other; and as those mining fans are comparatively large, the engines set on the fan shaft may be run fast enough to give the required speed. This saves the friction and expense caused by the use of belting; and this friction is considerable, when we take into consideration that extra friction is caused by an engine working on a separate shaft, together with the weight brought on the bearings by the pressure of a heavy belt running over the pulleys. Then there are the wear and tear of the belt, which form items important enough to require serious consideration.

To sum up, you will find that those fans having the largest inlets and outlets do the most work, according to their size, if the proper width of vane is provided;

That the passages leading to a fan should not be less than one-fourth in area in excess of the area of the inlets leading to the wheel, and that the inlets and outlets should be equal in area to each other;

That the projectile force of a fan-wheel depends on the velocity of the outer extremities of the vanes (it must be remembered, too, that any obstruction in front of those vanes will prevent the projection of air in direct proportion to the area of such obstruction);

That to restore the equilibrium of the air contained within the limits of the fan-wheel, gravitation forces air into the partial vacuum, in lines parallel with the revolving shaft, and this motion of the air will be manifested to the outer extremities of the vanes, provided that there is *no obstruction* placed in its path (if there are such obstructions made, then the effects of the wheel will be reduced in proportion to the area of such obstruction);

That, according to the present general method of construction, fans have such obstructions placed in the path of the air-currents passing through them by the erroneous construction of their cases;

That air-currents should not, if possible, be caused to travel in any of the main passages of a mine at a rate greater than ten feet per second, and to keep the currents within moderate speed the air-ways should be made as capacious as possible; at all turnings of an air-trunk or passage the area should be doubled.

Where air-currents carry explosive mixtures, a safety lamp should never be exposed to their action when they move faster than six feet per second; and for this reason, lights used near fans discharging such mixtures should be discarded.

Concerning the high velocities of air-currents or air-displacements caused by fan-wheels, experiments have shown in two remarkable instances that great force is needed.

By repeated trials with the same machine, it was found that it required ninety indicated horse power to displace 400,000 cubic feet of air per minute at a velocity of one hundred and twenty feet per second, through a passage of eight feet four inches which formed the inlet to an eight feet wheel open to the air on one side only. Another wheel, on the same plan, displaced 100,000 cubic feet of air per minute, at the rate of nearly sixty feet per second, through a passage of six feet two inches in diameter, with an expenditure of fifteen horse power.

In both instances the full force of the engines was required to produce the effects stated above, and the displacement was effected in the open air.

In the former instance the power was obtained from an engine having a cylinder of twenty-five inches diameter and twenty-one inches stroke, followed by a pressure of steam equal to forty pounds per square inch. The revolutions being forty-five per minute, the power was, therefore, about ninety horses, as shown.

In the second example the cylinder was of eight inches diameter and twelve inches stroke, and being run at the rate of seventy revolutions per minute, under a pressure of seventy pounds per inch, the power developed approximated fifteen horses.

The results of these experiments tend to shake one's faith in certain rules regarding the friction of air-currents or displacement of air. They show that the resistance to forces does not increase as the square of the velocity, and that the power necessary to promote air-currents does not increase as the cube of the velocity, and that certain statements made in some of our text-books, concerning this particular, do not accord with practical results, any more than they do in regard to floating bodies moving rapidly through water, such for instance as a large ocean steamer.

One reason may be advanced for this apparent variance. For example, a fan is set to work at a certain point. At its commencement it affects the particles of air in close contact with the space within which it revolves, and motion is communicated gradually to air, at points more remote. Thus a particle of air at some remote place is acted on and begins its motion towards the fan at the lowest conceivable rate. As it approaches the fan its motion becomes accelerated. By the time it has arrived at the point where the rarefaction is greatest, which is within the space in which the fan is revolving, it is invested with its full motion, and possesses a certain amount of momentum which has been imparted to it, in a manner so gradual as not to have affected the machine in the same way that it would have done had the motion been given to it suddenly. The motion of the infinite number of particles which compose the current when collected together to enter the fan-wheel, thus represents a momentum very similar to that of the fly-wheel of a steam engine when running at full speed. Were the fan-wheel stopped after a high velocity had been given to an air-current, the current would continue in motion, like the fly-wheel of a steam engine after the steam is shut off, until brought to rest by external causes.

Although we do not intend to enter fully or scientifically into the subject here, enough has been said to explain the mystery of the paradox to any one having a correct knowledge of mechanics as a science.

Therefore, it must be said, in conclusion, that the momentum of an air-current has much to do with the power that gives it motion, and acts in the same manner or on the same principle that a steamship does in the displacement of water lying in its path.

Thus, a steamer having an immersed cross section of 900 square feet, when running at fourteen knots an hour through the water, will be resisted by a force of not less than one hundred and eighty tons in the face of the finest ends that can be made to push the water aside in front as slowly as possible, and to allow it to flow into the vacancy constantly forming at the vessel's stern, as rapidly as those lines astern can be made to admit of such a desirable end, and prevent that *drag which retards the progress* of a vessel as much as does the displacement in advance of it. But in this case, although the push of the engines does not exceed, in many cases, the one-twentieth part of the resistance, the immense force accumulated in the vessel, in the shape of momentum, when taken into consideration, is sufficient to solve the riddle, and account for the errors made by Dr. Lardner in regard to this very subject.

At Usworth Colliery, in the county of Durham, England, a fan of forty-five feet in diameter is worked by a pair of engines, whose cylinders are each thirty-six inches in diameter by as many in length of stroke.

There are several fans in Pennsylvania where, when the diameter of the fanner is eighteen feet, the diameter of the cylinder on the main shaft is of as many inches, and length of stroke varying. But no reliable data have been furnished by the working of any of them that is sufficiently correct for insertion here; therefore, further particulars are not necessary.

There seems to be one error, very popularly entertained, concerning the working of the fan which must be pointed out before we leave this subject, as it seems to lead people astray when planning and constructing them for purposes of ventilation. It seems to be the opinion of some that if the air is revolved to a certain point, it will be thrown off with greater vigor at this point, on account of having by the action of the fan received an accelerated motion. This can never be so, where the air-passages are sufficiently large to keep up the supply of air to the fan as fast as it can be projected. The moment the air is inside of the fan space, it revolves as fast as the fan is driven, and if it were revolved till doomsday, it would never increase in speed, but would revolve with the fan, as if it were a component part of the machine, provided that the means of escape were shut off, or the means of ingress were closed. But the passages being free, the air passes off in a stream tangentwise, without following the direction of a vane a point farther than that which leads it out to the vane's extremities.

When the passages are too small, however, this will not be the case; because, on account of the air being inadequately supplied to a *good fan*, it will not only be made to revolve uselessly in the fan-wheel, but be caused to revolve by contact before it can enter the fan. This is caused by the drag occasioned by repeated contractions of the air-passages. When these cannot be enlarged, then in order to save power, the next best thing to be done is to reduce the inlet and outlet of the fan to correspond with such contractions of air-passage.

It may be asked, How it is that the fan working as it does within a space, to rarefy the air, throws its load off into an atmosphere where the pressure is somewhat in excess of the air entering the fans? In many cases, the weight of the external air is one-sixteenth greater bulk for bulk than the air coming up to the fan, and nothing is added by any power of compression to the air's density, after it enters the fanwheel.

To satisfy yourself on this point, take a piece of cork, and throw it into a body of still water. By its projectile force it will be driven below the surface of the water. This reduces the principle of fan-action to this: That the velocity and weight of the air in the fan, together, exceed the atmospheric pressure so much as to cause its displacement.

CHAPTER XXV.

UNDERGROUND FIRES, AND METHODS OF EXTINGUISHING THEM.

Fires have frequently been started in coal mines by various causes. The mines mostly subject to fires are those which contain large amounts of combustible gas. A stream of gas coming from a wall of coal has been lighted up by an explosion, and having been left to burn by the men, who have retreated to get out of harm's way, has got so far ahead as to defy all efforts made to extinguish it.

Before a fire is spread to any considerable extent, it may easily be put out by shutting off the air. In this case great judgment is required in order to perform it in a proper manner. Space must be left between the batteries and the fire, to allow the gases to circulate and distribute the heat generated at the seat of the fire; because soon after the stopping off of the air is accomplished, the air within the inclosed space will become so mixed up with gas as to be incapable of supporting combustion. This gaseous mixture will none the less circulate throughout the space, and the effect of the circulation will be to distribute the heat, and divide it among all the surfaces of the inclosed space. After this is accomplished, it is very likely that the fire will be extinguished. Where such fires have been closed up for a space of three months, it has been found that they were completely extinguished. But it is a dangerous operation to open out such a space, if there exists any spark of fire, or any heat so great as to start up a conflagration as soon as a circulation of fresh air takes place. It is not only a fire which has broken out at the opening out of such places, but in one instance, at Laffak Colliery, Lancashire, England, where the stoppings were opened which tapped a space in which there had been a fire, an explosion took place of a terrific nature. The two officers of the mine, who opened the space, were literally roasted alive.

The composition of a gaseous mixture forming in such an inclosed space can better be imagined than accurately described. After a certain time the air within the space will cease to give out any more of its oxygen to the fire, and the space will then be filled with a mixture which may be supposed to be largely made up of carbonic acid, and, not unlikely, with carburetted hydrogen also. How such a mass coming out into the fresh air will deport itself in the presence of a safety lamp it is difficult to say. Such masses should be let out gradually by the opening of small holes through the stoppings, or if no more than one stopping be used, one hole made at the bottom and another at the top of it, would in a day or two allow the gases inclosed within the space to exchange volumes with the air on the outside. This was done at a colliery near St. Helen's, Lancashire. The fire had been closed up during three months before small holes were made through two of the stoppings. This allowed an exchange to take place between the air of the mine and the gases of the inclosed space; but this was done in a manner so gradual, that no danger or inconvenience from the gaseous mixtures was encountered. When, at the end of two weeks after the first small openings were made when the stoppings were taken down, it was found that the fire was not only out, but the place was cool enough to be entered and reopened to the workmen.

But the method would fail in the hands of any one, unless due regard were paid to the manner of arranging the stoppings, so as to allow of a certain amount of space for the circulation of the inclosed air and gases; and the time of its being closed up would need to be proportioned to the time of burning and other circumstances connected with that of the fire.

In case of fires breaking out in any mine that is open or owing to the seams cropping out to the surface, the shutting off of the air is not so easily accomplished. There is reason to think that in shallow mines where a large fire is burning, by reason of the caving of the roof into the vacancy made by the fire as the coal is being consumed, connection is made with the surface through crevices which supply air to the burning masses.

If the fire is very extensive, as it will be if the coal seam is large, and if the fire has been in progress a considerable length of time, in order to confine it within certain limits, the coal should be mined out as speedily as possible around the pillar in which the fire is operating. When thus isolated by a breach in the strata, which the mining should be made to form, the fire cannot extend further than the fallen masses. But when this is the case, the large amount of gas formed by the fire must be allowed to escape to the surface; and if the ventilating currents are properly stopped off from the burning district, there will be no danger encountered by working in the other portions of the mine.

Water has been run into mines, and fires have been successfully extinguished through its agency; and where the mines are below water level, and water is convenient, it must be the most reliable and speedy of any method known.

By the forcing in of gas, or a mixture of gas and vapor and air so as to form a mass that is not capable of supporting combustion, a fire may be extinguished at once. But a large body of the gaseous mixture is required to be kept in constant circulation past the seat of the fire, for several months together, and without any intermission; because, on account of the heat occasioned by the fire, which heat must be literally carried out of the mine by the current of gas, the gas must be kept in circulation for a sufficient length of time to reduce the heat so much below the point of combustion that there will be no possibility of the fire re-commencing as soon as the fresh air is admitted to the mine.

Gas may be made very cheaply for this purpose by the combustion of coal refuse. If furnaces were erected at once at the mouth of a mine, and a few boilers laid over these, steam could be raised to work an engine and fan. The fan used to force air through the furnaces under a pressure of about three inches of water, would force the gas of the furnaces and steam of the engine through the mine, under a pressure sufficient to prevent the influx of air at any other point; provided that all other openings were so reduced in area as to cause the streams of escaping gas to be driven out of the mine under a pressure of two or three inches of water, or a pressure above that of the external air.

The cooling of the gas will be most easily effected if it should be forced through a body of water in a tank, or in a dam made near the entrance of the mine. The hot gases passing through a body of water or through a cloud of spray would, in cooling, evaporate a certain amount of this water or spray and mingling with it would form a mass incapable of supporting combustion. If forced past the fire, the heat would be carried off by convection. Provision to carry the current to the required point should be made previously to turning on the stream of gas, which should not be allowed to cease until the fire is extinguished. The condition of the mine could be judged of by the daily application of thermometers to the escaping currents of gases.

Except that gas could be generated by the burning of refuse coal, it could not be produced by other means in sufficient quantity to perform the double office of extinguishing the fire and reducing the heat.

In many cases the steam-boilers and furnaces of a colliery may be employed to perform the above-named duty, and the ventilating fan of the mine may also be made to force the gases into the fire.

When a fire started by a stream of gas coming from a crevice has burnt long enough to cause great heat, carbonic oxide gas is generated. When this gas is present, a fire burning at the crevice may have been dashed out completely; but if a certain amount of heat is spread about the locality, and fresh air is allowed to mingle with the carbonic oxide, an explosion will by their contact ensue; and you will see repetitions of the phenomenon at the top of a heating or puddling furnace in a rolling mill, when such furnace is in full blast. This gas seems to require nothing more than heat and fresh air (without flame), to start a conflagration. Who knows that this gas has not been the cause of serious mischief in mines by being generated from coal dust burning within the wire cylinder gauze of a safety-lamp? Are not there cases in which this gas, when made by the burning of a lamp, may have been the cause of igniting an explosive body on the outside of the lamp's gauze?

What of the accident at Avondale, Pa., where a body of heated gas came off a furnace, and ascended a shaft to a drift through which fresh air could have been admitted? We know that the result was an *explosion*, followed by a conflagration, and its fatal consequences!

CHAPTER XXVI.

A GENERAL DESCRIPTION OF THE COAL VEINS WORKED AT ELLENWOOD COLLIERY, SITUATED IN THE SOUTHEASTERN BRANCH OF THE MAHANOY COAL BASIN, GIVEN TO SHOW THE GREAT NATURAL RESOURCES OF THE ANTHRACITE COAL FIELDS.

At a point southwest of Ashland, on the flank of the Mahanoy Mountain, where the coal strata come up to the surface inclining from the horizon at the rate of 50°, they are interrupted by the formation of an anticlinal dip. From this point an anticlinal axis starts to the eastward, and its main centre runs about two and a half points of the compass to the north, while the main anticlinal axis of the Mahanoy Mountain runs about one point south of east. About four miles to the eastward, the anticlinal axis which starts out from the point specified above, has carried away with it so much of the strata as to form the Bear Ridge Mountain. But from the same point of departure of the anticlinal axis, a synclinal axis also starts out, whose general direction is nearly one point to the north of east. This synclinal axis forms the southeastern limit of the Mahanoy Coal Basin, and it forms a small basin of a dozen miles in length, whose breadth at Mahanoy City is not less than three-quarters of a mile. It is in this valley that the town of Mahanoy City is built, and the valley carries a branch of the Mahanoy Creek, which drains the water from the adjacent mountains, which are the Mahanoy to the south, and the Bear Ridge, or one of its projecting arms, to the north, whose ridges rise a few hundred feet above the level of the valley.

The coal seams below the Mammoth, as well as the Mammoth itself, crop out above the level of the creek, one after the other in the same order in which they lie in the strata; and they rise to their out-crops at high angles of inclination, beginning at zero in the centre of the basin, and bending upwards gradually at first, as we may reasonably suppose from visible indications, until they assume the high angle of "dip" by which they present themselves to the light of day; 60° being the north dip and about 40° the south.

GENERAL DESCRIPTION OF COAL VEINS WORKED AT ELLENWOOD COLLIERY.   167

In spite of the immense forces to which the strata have been subjected in order to bend them into the U shape in which they are now found, the coal veins are in a fair condition, with their roofs and floors in tolerable conformity with each other.

As a specimen of one of the collieries worked in the Mahanoy Basin, we have selected the Ellenwood only for the purpose of giving an example of the natural resources of the basin in respect to the coal seams.

Breaking ground on the flank of the Bear Ridge Mountain, which forms the northern limit of this branch of the basin, we have at this point a thin coal seam opened, the Skidmore, which lies under the Mammoth Vein at a distance of from twenty to twenty-four feet. At present this seam of coal is not considered workable on account of its being mixed with impurities; but as it dips in perfect conformity with the strata inclosing it, its course is followed to the south by a sloping shaft to a depth of about seven hundred feet on an incline varying from 30° to 45°.

By excavating in the seam of coal to a sufficient width, and by blasting down the roof in one place and the bottom or floor up in another, a height and width have been made sufficiently large for the working of the slope wagons, which carry upwards of two and a half tons of coal when properly loaded. Besides there is room made on one side of the slope for the pump used to drain the mines, one of Joseph Allison's double plunger steam cataract variety.

At the bottom of the slope a short tunnel opens a road to the Mammoth Vein, or at least to the under portion of it, as at this point the conglomeration of veins, usually known as the Mammoth farther to the westward, is separated into two distinct veins, each of which is worked by independent mining operations.

At the point where the tunnel mentioned above intersects the vein, the main roads or gangways branch off, one being driven to the east, the other to the west. In this under portion of the Mammoth a breast of coal shows us a very beautiful section which measures in thickness twelve feet in some places to sixteen feet in others, and is divided into three well-defined benches. The slates dividing the benches, together with the streaks of "bone coal" or "splints" and sulphurets, do not possess more than eight per cent. of the vein's space, ninety-two per cent. being coal of the best quality of anthracite, which is so compact, that when two pieces are struck against each other, they emit a metallic sound.

From the line of gangway driven in this under part of the vein, tunnels are driven through the slate which separates the upper part of the seam from the lower. This section of the vein gives us in thickness of coal about twenty feet. The body of slate which divides the Mammoth Vein is here about fifty feet thick, and it forms the roof of the under portion of the seam, and the floor of the upper part of it. As the coal is worked out according to the popular method of getting coal in the Schuylkill region, which has been already fully described in detail in former chapters of this work, we need only refer for details to Plates XIX. to XXIV. Putting the two parts of the vein together, we have a mass of coal, the magnificence of which must excite the admiration of any one who has not become familiar with the coal measures of this part of the world. But this is not all of the coal which is worked at this colliery! Leaving what may be above the Mammoth in the Primrose and other veins which are not opened or worked here, by tunnelling north through the strata two more workable veins are cut, the first of which varies in thickness from seven to nine feet; while the second is the Buck Mountain, whose thickness averages about sixteen feet. In all there are masses of coal opened whose aggregate thickness amounts to sixty feet.

There is an old rule, which has been much used by coal mining engineers in estimating the value of coal lands, which if applied to the coal lands composing this Mahanoy Valley, would place them at a high standard. This rule allows one thousand tons of coal per acre for each foot in thickness contained in the ground. This, in the English and other European coal fields, makes ample allowance for waste produce, and for the faulty portions of a mine's property. Then, at this rate, there must be not less than between fifty and sixty thousand tons of coal to each acre measured on the inclined planes of the coal strata. Of course this large amount of coal is within that portion of the basin bound by the out-crops of the Mammoth Vein. But the out-crops of the underlying veins run farther up the mountain sides, and in consequence of this, their area is extended in a ratio to correspond.

The above description has been obtained from a personal inspection of the veins being worked at the Ellenwood Colliery; and this colliery has been selected at random from about fifty which are in the Mahanoy Basin, the majority being worked by the Philadelphia and Reading Coal and Iron Company, which is without exception and by far the most extensive coal mining company in the world.

The coal lands in the possession of this company, some of which are owned entirely, others partially, while others are leased by this company, are so vast as to warrant a heavy freightage for the Philadelphia and Reading Railroad, a sister corporation, for a century to come; and on this account this road can never be rivalled in the carrying of coal as long as it holds on to a controlling interest in the mining of it. It was by a master-stroke of policy that Franklin B. Gowen conceived the idea of organizing this coal company, which has become, by dint of his perseverance and much hard mental labor, embodied into such vast proportions.

A branch of the Philadelphia and Reading Railroad threads its way among the valleys of the Mahanoy Basin, and gathers from the mines not less than twenty thousand tons of coal daily. This amount of coal finds its chief outlets over the inclined planes at Gordon, which run up the slopes of the Broad Mountain, and over the Mahanoy planes near Mahanoy City, and by way of the tunnel through the Mahanoy Ridge at the eastern extremity of the basin. Most of the coal is carried by the main line to Philadelphia, and it forms only a portion of the whole of the coal carried by this company.

The Northern Central Railway, a branch of the Pennsylvania Railroad, taps the Mahanoy Basin at its western extremities, and carries much of the basin's produce to the Baltimore markets, and a branch of the Lehigh Valley Railroad enters by mountain passes at a northeastern point and carries away daily out of its valleys several thousands of tons.

INDEX.

Abandoned excavations, examination of air in, 12
Accidents in mines, prevalence of, in early times, 82, 83
    means to be used as safeguards against, 9–13
Air and gases, expansion of, 82
Air courses, driving, 13
    should be straight and direct, 13
Air crossing, 12, 63
    in ventilation, 39
Air crossings, 36
    driving in solid strata, 23
Air currents, velocity of, experiments with, 152
Air doors, 13
    worked by gravity, 63
Air in abandoned excavations, examination of, 12
Air, splitting the, 36
Air stoppings, 13, 36
Air-way, spacious, 12
America, mining in, 59
    want of experienced miners in, 75
    waste of coal in, 74, 75
Anemometer, the, 43, 45
Angel of death in the air in mines, 13
Ashland, Pa., colliery at, 166
Avondale, Pa., explosion at, 165

Baldwin Locomotive Works, 144
Battery and loading platform, 104, 105
Battery collar, 105
Bear Ridge, 166, 167, 169
Belgium, mines of, 59
Bell, Isaac Lowthian, 59
Blackboard, putter's, 65
    the deputy's, 48
Black Fell colliery, 78
Blasting coal, 107, 108
Board and pillar and long wall systems of mining compared, 142–147

Board and pillar system of mining, 60
    system of mining out coal by, 33–88
        some advantages in, 74
        with experienced miners, 74, 75
    whole coal working, plan of, 36
    workings, ideal plan of, 37
Board and wall system of working, 37
Board room, interior view of a, 88
"Board-ways," 37
Boards and ends of coal, 36–41
Bore holes in testing for gas in mining, 12
Bonses in coal mines, 26, 27
Bottom rock, gutters in, advantages of, 128–131
Buck Mountain vein, 90
Buddle, John, improvements in ventilation, 83
Brattice in an opening schute, effect of, in ventilating, 21–23
Brattice in Europe since the time of Spreding, 23
Breast of a mine, 106
Breast-rooms, 104
    and pillars of a mine, 103, 104
Breasts in the Mammoth vein, working, 113–119
Broken coal worked, 63–70
    working in the, 76
Broken workings, details of, 71, 72
Brushing out the gas in a mine, 21

Cages used in a Newcastle shaft, 33, 34
"Calling course," 85
Carburetted hydrogen, detecting in the air, 13
Centrifugal fan, 150
Cleavage of coal, 137
    or "grain" of coal, 115
Coal, broken, worked, 63–70
    cleavage of, 37
    deposits, 98–101

(171)

Coal-dust and coal-gas in the air, a dangerous mixture, 14–16
Coal formations, 98–101
  from the highly inclining coal veins of the United States, how taken, 89–147
Coal-gas and coal-dust in the air, a dangerous mixture, 14–16
Coal, immense waste of in America, 74, 75
Coal lands of the Philadelphia and Reading Coal and Iron Co., 166–169
Coal mined, its relation to the ventilation of a mine, importance of, 28
Coal miner and his work, 54–57
Coal mines and coal miners, general information concerning, with description of long wall as worked in horizontal seams of coal in England and France, 9–32
Coal mines as worked in the thick coal measures, plan of, 103, 104
Coal-mining engineers, 24
  in France by Remblais, 29
Coal seam, rule for the yield of per acre, 29
Coal seams in the Newcastle coal field, 33
  of Staffordshire, England, 29
  thick, 29–32
Coal veins at Ellenwood, 166
  highly inclining, 89–147
  in France, 29
Colliery near Mahanoy City, Pa., described, 166–169
Comparison of the different systems of mining, 142–147
Comparisons and remarks, 73–77
Continental mining, lessons to be derived from, 59
Conversation on the principles of ventilation, 120
Conversations with miners, 90–96
Cooper and Hewitt, 60
Cooper, Peter, 60
Corf, 81
Crawshay, Mr., 60
Creep, bringing on a, 73, 74
" Creep," keeping off a, 80
Creeping down of the roof of a mine, 73, 74
  in of the bottom slate, 76
Crossing the aerial ridges of the board rooms, 76
Currents of air in mines, 83

Death, angel of, in the air of mines, 13
Deputy, the, 45
Deputy's experience, a, 42–46
Derbyshire, long wall of, 60

Details in working and ventilating a district, 42–46
  of broken workings, 71, 72
District and panel workings, 36–41
  details of working, 36–41
Dividing the ventilating current, 12
Drawing of a jud, 67, 68
  out the props, 67
Drift, mouth of, 102
Drilling coal, 109, 110
Drill, the, 106, 107
Driving a breast, 113–119

Early collieries, reworking, 78
Ellenwood Colliery, Mahanoy City, Pa., description of, 166–169
" Endways," 37
Engine bank, 61
Engineer, the educated, a necessity for, in coal mining, 27
England, fan in use in, 160
English coal measures, shaft through, 33–35
  mines, early ventilation in, 83
  mining, changes in, 60
Examination of breasts, 134–141
Excavations, abandoned, examination of air in, 12
Experience of a deputy, 42–46
Experiment with marsh gas, 17–19
Explosion and early reminiscences regarding, 9–11
  at Avondale, Pa., 165
  in Springwell Colliery, 9–13
Extinguishing underground fires, 162–165

Fan action, manner of, 154–157
  at Usworth Colliery, 160
  boys, 20
  dispensing with, 20
Fan wheel, operation of, 151
Fans in use in Pennsylvania, 160
Fatfield Colliery, 78
Field, Cyrus W., 60
Fire at Laffak Colliery, 162
Fires underground, extinguishing, 162–165
Flange of the rail and of the wheel of the railway wagon, 82
Flint mill in mines, 82
France, coal veins of, 29
  mines of, 59
  mining in, by Remblais, 143
  working thick, highly inclining coal beds in, 89

French system of mining, safety in, 143
Friction of fans, prevention of, 157
Furnaces for ventilation, early, 83
Furnace, ventilating, 82, 149

Gangway timbers, 91, 92
Gangways, 104
Gas and air, expansion of, 82
Gas, extinguishment of fires in collieries by, 164, 165
   in a coal mine, causes of accumulation of, 42
      watching the accumulation of, 42–44
Gas, testing for by bore holes in mining, 12
Gases, absence of a knowledge of in early times, 82
   in mines, how detected, 43, 44
Gateshead Iron Works, 60
General conclusions, with a comparison of the different systems of mining, 142–147
General information concerning coal mines and coal miners, with description of long wall as worked in horizontal seams of coal in England and in France, 9–32
General remarks, 81–88
Germany, mines of, 59
Goaf, ventilating a, 73
Gowen, Franklin B., 169
Gutters for drainage in the bottom rock, advantages of, 128–131
Gutters for drainage of mines, position of, 128–132
   in the coal, disadvantage of, 128–130

Haswell Colliery, 47
Headley's treatise on coal mining, 27
"Headways," 37
Hetton Collieries, 24–28
   gigantic works at, 25
Hewer, the, and his work, 54–57
Highly inclining coal beds of France, working, 80
   coal veins of the United States, taking the coal from, 89–147
Hoisting machinery in mining, 27
Hoisting ropes, inspection of, 13
Horse roads in mines, 35
Horses, getting into the level on trucks run down the engine plane, 35
Hutton seam, Newcastle coal field, 33–35

Improvement of mining in the Southern and Webb mines, 24–28
Incline bank, 64

Inclined plane or engine bank, 35
   selfacting, 58–62
Inspecting mines daily, 13

Jud, drawing of a, 67, 68
Juds, working off the, 67
Juggler manway, 113

Laffak Colliery, Lancashire, Eng., fire at, 162
Lagging and timbering a mine, 102
Lampton Colliery, 78
Lancashire and Newcastle coal fields, 80
   collieries, improvements adopted at, 26
   mining in, 24
   modification of board and pillar system in, 80
   panel system at, 27
Lardner, Dr. D., 160
Lehigh Valley Railway, 169
Loaders, 110–112
Long wall and board and pillar systems of mining, 142–147
Long wall mining in America, 75
   some of the advantages of working by, 74
   system of mining in Derbyshire, 60
   in Staffordshire, 30–32
   system of working, 37
   working coal by, in England, 29

Mahanoy basin, 169
   city, 166, 169
   coal basin, colliery in, 166
   coal region, 89, 90
   creek, 166
   mountain, 166
   coal strata of, 90
   planes, 169
   Valley, 114, 168
Mammoth coal vein, 29, 90, 114, 115
Manway door, 113
Manways, 113–119
Marsh gas, an experiment, 17–19
   procuring for experimental purposes, 19
Master shifter, 58
   wasteman, 58
Measurement of work done by miners, 128
Measuring of work, 136, 137
Metal rigs and old coal pillars, 79–80
   driving through in old excavations, 76

Miners and their bosses, prejudices of, 19-23
    conversations with, 20-26
    tools, 106-110
Mines should be inspected daily, 13
Mining appliances, the older, 81
    coal in seams of less than seven feet, 80
    comparison of different systems of, 142-147
    importance of organizing the men and dividing the work in, 144
    in the Southern and Welsh mines, improvements of, 24-28
    of coal, details of, 102-112
    out coal by the board and pillar system, 33-88
Monkey gangway, 104

Needle, the, 106, 107
Newcastle and Lancashire coal fields, 80
    coal field, Newcastle, England, coal seams in, 33
    the men and the mines in, 81-88
    early accidents in the collieries at, 83
    system of organization of mining in, 144
Northern Central Railway, 169

Officers of mines should practise the art of discovering the presence of gas in air, 13
Old coal pillars, reworking, 78-80
    mines, reworking, 78-80
    places, examination of air in, 12
"Onsetter," the, 33
Organization of an improved English coal mine, 27
Overman, the, 58-60
    the duties of, 58
    the importance of, 58
Overman's cabin, 81
    tracing, description of, 36

Panel and district workings, 36-41
    of coal, time required to work off, 79
Panels of coal in Lancashire coal mines, 79
Parleys with miners, 135-141
Pennsylvania coal district, topographical features of a, 80
Pennsylvania, mammoth coal vein in, 29
    railroad, 169
Philadelphia and Reading Coal and Iron Co.'s collieries, 166-169
Pick, 108, 109

Pillars, 115
    and breast rooms of a mine, 103, 104
    millions of tons of coal wasted by, 74
    splitting the, 36, 76
Prejudices of miners and their bosses, 19-23
Props, dispensing with the use of long, 74
    drawing, 74
    drawing out the, 67
Plans of workings, with descriptions, 36-41
Props, weight of the roof on the jud coming on the, 67
"Putter," origin of the, 81
    the, 47-53
Putter's blackboard, 65
    work, the, 48

Railways, foundation of, 81
    on the surface, origin of, 82
Rail, wooden, origin of, in the Newcastle coal field, 81
Regulating doors, 12
Regulator, the, and air crossing, 63
    the, in ventilation, 40, 41
Remarks and comparisons, 73-77
Renablais system of mining in France, 143
    working out coal by, in France, 29
Rents in America, paid by the ton, 74, 75
Reworking of old mines, metal rigs and old coal pillars, 78-80
Rise side workings, 36
Roof, crushing of, in long wall mining, 31
    getting down the, when the props are all out, 68
    of mine, creeping down of, 73, 74
Ravensworth Colliery, 78

Safeguards against accidents, means to be used, 9-13
Safety in mining by the French system, 143
    lamps, 82
    use of in the broken coal, 69
Safety lamps, attention to, 13
Schute, 105, 106
Schuylkill County, Pa., coal formation in, 89
Scraper, the, 107
Screw fan, 130
Self-acting incline plane, 58-62
Shaft through English coal measures, 33-35
    the Hutton coal seam, Newcastle, Eng., 33
Shenandoah Valley, Pa., 114
Sledge, 109
Smyth, Warington, 154

Spedding's application of underground ventilation, 23
   improvements in ventilation, 83
Splitting the air, 36
   the main current of air in coal mining, 39, 41
   the pillars, 36
"Splits," 12
Springwell Colliery, 9, 83
   explosion at, 9–13
Squib, a, 108
Staffordshire, coal mining by long wall in, 30–32
   England, the coal seams of, 29
   thick seam of coal, mining by long wall, 32
Starter's battery and loading platform, 104
Starter, the, 105

Tamping for a blast, 108
Technical school of the mining engineer, 84
Thick coal seams, 29–32
   seam of Staffordshire, working, 32
Timbering a mine, 102
   examination of, 13
   the manways in the breast-rooms, 113
Topographical features of a Pennsylvania coal district, 89
Tracing, overman's, description of, 36
Tram, origin of the, 81
Tramway, origin of, 81, 82
   used in Durham and Northumberland, 64
Trapper, the, 63
"Trip," the, 90
Tubs used in a Newcastle shaft, 33
"Turn," the invention of, 81

Underground fires, and methods of extinguishing them, 162–165
United States, highly inclining coal veins of, how coal is taken from the, 89–147
Upheavals, 98
Usworth Colliery, Durham, 160

Ventilating and working a district, details of, 42–46
   a goaf, 73
   by means of a brattice, 21–23

Ventilating—
   current, dividing the, 12
   fan, how it should be constructed and arranged, and the principles of its action, 149–161
   fans, the respective qualities of, 150
   force, mechanical, 13
   furnace, 82, 149
Ventilation, defective, 135, 136
   details of, 120
   early furnaces for, 83
   in English mines, early, 83
   of coal mines by splitting the main current of air, 39, 41
   precautions to be taken in regard to, 12
   progress of improvement in, 83
   regulated by the quantity of coal mined, 28
   relation of, to the amount of coal cut in a given time, 145

Wales, John, 24–26
"Wall's end" or lump coal, 32
Washington Colliery, 78, 83
Waste of coal in America, 74, 75
Water-course, 102, 128–131
Wedge, 109
Weight of the roof in the jads coming on the props, 67
Welsh collieries, improved system adopted at, 26
Westphalia, Belgium, and France, mines of, 59
White damp, 135
Winding in shaft, 34
Wood, Nicholas, founder of the Northern Institute of Mining Engineers, 24
Working by crossing the metal ridges of the old board rooms, 76
   and ventilating district, details of, 42–46
   levels, 36–41
   out coal in France, 29
   through the metal rigs, 76
Workings, plans of, with descriptions of, 36–41
Wyoming Valley, Pa., mining in, 143

Yield of a coal seam per acre, rule for, 29
Yorkshire, system of mining in, 143

www.ingramcontent.com/pod-product-compliance
Lightning Source LLC
Chambersburg PA
CBHW031832230426
43669CB00009B/1324